CBers' Factbook

CBers' Factbook

NOEL T. SMITH

Research Institute for Advanced Technology
Killeen, Texas

HAYDEN BOOK COMPANY, INC.
Rochelle Park, New Jersey

To Judy
My partner in this and all endeavors

Library of Congress Catalog in

Library of Congress Cataloging in Publication Data

Smith, Noel T
   CBers' factbook.

   Includes index.
   1. Citizens band radio. I. Title.
TK6570 C5S555   621.3845'4   77-15036
ISBN 0-8104-0787-6

Printed in the United States of America

1	2	3	4	5	6	7	8	9	PRINTING
77	78	79	80	81	82	83	84	85	YEAR

Preface

Citizens Band Radio is a hobby filled with fun and excitement. It is a service of infinite utility. The proof of these statements is evidenced in the burgeoning numbers of individuals joining the ranks of its users. Few activities have stirred such an enthusiastic response and fewer still have continued to generate ever greater interest with the passage of time as has CB radio. The explanation must be more than mere faddism. Indeed, CB radio seems destined to become an American institution.

This book is intended as both an introduction for the beginning CBer and as a guide book for the experienced. The topics are practical in nature. Much of the material is in a "how to" style. The CBer who wants to "do it himself" will find the discussions and illustrations helpful. The person who chooses to pay someone else to install and adjust his radio will depend on this volume as a guide in evaluating the work of those they hire.

CB radio divides into two basic categories, base and mobile. A base station is fixed in location. It usually utilizes standard ac power and tower-mounted antennas. Because room at a base station is less of a problem than with mobile installations, accessories are more abundant. Base station operation is discussed in Chapter 2.

Mobile operation is highly popular among CBers. The fact that travel is often monotonous and boring is undoubtedly a contributing factor to this popularity. The ability to keep in touch and to render aid to stranded motorists and at accidents also prompts people to install CB radios in their cars. Chapter 3 deals with mobile operation and installation.

Antennas are the subject of Chapter 4. A radio is only as good as its radiated power. A very important part of the CB system is thus the antenna. Base antennas and antenna installations are discussed.

FCC regulations limit CB equipment specifications. Most units will provide signals of similar strength and quality. If a person desires to be outstanding in his communications ability, he will have to seek ways of improving his signal quality above that of his fellow CBers. Chapter 5 describes ways of improving CB communications by operating techniques and technical means.

Accessories for CB radio are many and varied. Practically any device the CBer may desire is available as an accessory. Chapter 6 describes a few that CBers frequently use. Accessories which increase operating convenience and safeguard radios from theft are discussed.

My research into CB radio has been an enjoyable quest. Nothing contributed more to this enjoyment than the meeting of the friendly people along the CB trails. I offer thanks to those who shared their experiences in CB radio—Franc Wilson and Silver Fox and all the others in the Killeen CB Club. To Killer and Sparkplug, who received a 5 by 9 on our Hammond organ, I say thanks. P.M. and John at Mountain Top Communications were especially informative. My thanks also to the manufacturers who provided material for this book.

No book is the product of one person alone. This one has been enriched by the talents of Mrs. Lynne Ward and Jack Mabe. Their help in typing and proofreading is appreciated. A special thanks must go to Judy, who brought to my fumbling prose a glimmer of polish.

I hope you who read the book will find it an enticement to explore CB radio more fully. The fun and excitement waits only on your press of the microphone thumb switch. Happy CBing!

NOEL T. SMITH

Contents

CHAPTER 1

An Introduction to Citizens Band Communications

On a dark and lonely country road, an elderly woman notices the telltale thump of a tire going flat. She stops her car and looks about for a nearby house. All she sees is darkness! Her health and strength will not permit her to change the tire. What can she do? Where will she get help?

This woman is prepared for the emergency. She slips into the car seat once again and turns a knob until the number 9 shines out into the darkness. She picks up a microphone and presses the switch on it. Her thin, tense voice is heard, "Hello Channel 9 monitor; this is KZZ 1073 mobile. I need a 10-33.* Can you assist me? Over."

Another voice, high pitched with a mechanical sound, intrudes upon the stillness of the night. "KZZ 1073, this is KXX 2774. I read you loud and clear. What is your trouble, and what is your 10-20? Over."

"KXX 2774 this is KZZ 1073. I'm located about four miles north of the Jourdentown turnoff on Farm Road 106. My tire is flat, and I cannot change it. Over."

"10-4, KZZ 1073. I'm about two miles from your 10-20. I'll be there shortly. KXX 2774, clear."

"Thank you KXX 2774. KZZ 1073, clear."

This lady just experienced a use of Citizens Band Radio which is repeated many times every day. In fact, the use of CB radio for emergency communications is so extensive that the FCC has set aside Channel 9 to be used exclusively for emergency communications. The CB radio shown in Fig. 1-1 employs a briefcase portability which permits rapid response to any emergency situation. Highway emergencies, severe weather, sudden illness—many are the emergencies which benefit from communications on the Citizens Band.

CB radio is more than emergency radio, however. The local plumbing contractor keeps in touch with his trucks by CB radio. A wife calls her

---

*"10" signals are explained in Chapter 5. CB call letters used in this volume are fictitious, and any similarity to active calls is purely coincidental.

Fig. 1-1 *Security in a Suitcase:* Many people depend on CB radio for security while driving. Help is as close as the microphone. (*Courtesy*, Kris, Inc.)

husband to tell him to pick up bread on the way home from work. These are practical uses of CB radio. For such people and millions of others, CB radio is a very convenient way to stay in touch.

CB radio does not stop with the practical, though. It is rapidly becoming a major hobby. An orange pickup roars to a stop in response to the red glow of the traffic light. Shifting gears with one hand, clutching a microphone with the other, "Killer's" rapid chatter can be heard:

"We thank you, there, Mr. Sparkplug, for the break. We shore do, mercy sakes, you cotton picker. Mercy sakes, we definitely do. That's a forty roger on your load of Volkswagen radiators, golly, it shore is. We hope you fill 'em up real soon, mercy sakes, we shore do. Well, we've got to back 'em on out here, mercy sakes, we shore have. Golly, hope we get to see you in a short short for an eye ball to eye ball, mercy sakes, we shore do. Well, Mr. Sparkplug, bring 'em on back one more time, then the Killer has got to go, bye-bye. . . ."

Conversations like this one illustrate a new art form, CB slang.* As spoken, it becomes a chant, with meter and rhyme as set and patterned as in poetry. This is CB radio at its most enjoyable. This is CB radio as fun for fun's sake. No one worries because communications efficiency is reduced. The talk is for the sake of talking, and it is enjoyed by the listener as much as by the talker.

CB radio is a many faceted medium. Within one of these facets is something for you. What it is depends in large measure on your personality and how you choose to use CB communications. Regardless of your choice, CB radio stands ready to meet your needs.

CB Psychology

The recent popularity of CB radio is due in large measure to things which the CBers themselves might find difficult to identify. CB radio as it is currently being used fulfills a need to belong. The "good buddies" of the air always make room for one more. "Breaker, I hear you. Come on." People of different backgrounds and clashing personalities all belong to the same group. They are isolated from close scrutiny behind the microphones and antennas, but they are united through the incessant sounds coming from their speakers.

The popularity of "handles" and slang are all part of the phenomenon. Mr. Milquetoast can make big claims, and the staid Miss can afford a flirtation. CB radio fosters a psychology which could be called "anonymous exhibitionism." Names are changed to protect the innocence of the innocent.

Some users of CB radio are not so innocent. "Smokey Patrols" check roads and relay radar information. Those who choose to speed react to the "Smokey Report" by "letting down the hammer." As often as not, a few miles down the road will be Smokey the Bear, "ears" and all. A surprising amount of CB slang relates to speeding. The driver speeds up based on an "all clear" by some anonymous CBer. The comradery in this case becomes a bit nefarious. The psychology derives from a widespread tendency to want to "beat the system" just a little.

Citizens Band radio is based on humanitarian psychology, too. A large number of CBers are engaged in emergency and assistance operations. The operation may be of an unofficial and personal nature, or it may be a more organized concern on the part of a group dedicated to lending help. Radio Emergency Action Citizens Teams (REACT) operate to assist

---

*Appendix A contains a selection of popular slang words and phrases.

motorists in distress. The organization is voluntary and often provides 24-hour monitoring of emergency channel 9. A CBer interested in learning more about REACT can contact a local REACT team or can write to the national headquarters at the following address:

REACT National Headquarters
111 East Wacker Drive
Chicago, Ill. 60601

ALERT (Affiliated League of Emergency Radio Teams) is another public service CB organization. Somewhat similar to REACT, ALERT has a larger scope of activity, working in a larger variety of public situations. The address of the national headquarters is:

Affiliated League of Emergency Radio Teams
National Press Building
Washington, D.C. 20004

Some CB psychology requires no psychoanalysis. CB radio is just plain fun. CB chatter whittles down the miles of a long automobile trip and adds zest to a dull winter afternoon when indoors is the only place to be. A CB radio club can also be fun. Discussion is lively, and fellowship is abundant. Tall tales and CB legends are born and grow as neighbors of the air meet face-to-face. CB outings and jamborees are numerous. They are usually family affairs with different kinds of contests and activities on the program. CB equipment is often given as door prizes. At national and regional gatherings, impressive displays of CB equipment will be provided by the manufacturers.

CB radio has a potential for almost everyone. The only prerequisite is a willingness to communicate. All the rest will take care of itself.

The Evolution of the Citizens Radio Service

In 1947 the Federal Communications Commission established the Citizens Radio Service. The initial CB channels were allocated in the 460-470 megaHertz ultra high frequency band. The difficulties and limitations imposed by this ultra high frequency band kept communications short, and widespread public interest in it was nonexistent. Later, the number of channels within the band was reduced. The demanding requirements of Class A and Class B operation further limited the popularity of the service. This consequence caused no serious concern since the service was conceived as a practical use of radio communications in contrast to the hobby-oriented and experimental character of the Amateur Radio Service.

The Citizens Radio Service was expected to be used by businessmen and professionals who needed convenient contact with their offices and homes and by such service organizations as volunteer fire fighting units.

Use by the general public was not anticipated. One reason that the service was not considered attractive to the everyday citizen was the cost of equipment. Technology was young. Mass production had not come to radio. Vacuum tubes required sets of relatively massive bulk and complicated construction. In short, the time was not yet ripe for the Citizens Band, as it is now known.

In 1958, the FCC responded to an appeal by a growing number of citizens for a radio service which did not require license exams and utilized realistically priced equipment. Class D Citizens Band was born. At the time, the Amateur Radio Service had access to the 11 meter band. Because of this band's industrial users and unstable transmission conditions, the Radio Amateurs did not fully exploit its capabilities. The FCC, therefore, chose the 11 meter band as the location for a new radio service, the Class D Citizens Band Service.

This frequency band provided 23 channels. Equipment requirements were reduced. The day of the inexpensive CB radio was at hand. The subsequent development of transistors and integrated circuitry with the attendant economy of printed wiring and mass production reduced the size and cost of CB units while increasing their reliability.

The middle seventies introduced a milieu made for CB expansion. The citizenry was highly mobile. Population statistics were increasing. A psychology was developing which fostered "anonymous exhibitionism." Handles and rachet-jawing became the order of the day.

For several years, these relaxed operating procedures flourished without official authorization. Then, on July 29, 1975, the FCC adopted several amendments to the CB Rules and Regulations which officially designated the Class D Citizens Band as a hobby radio service. This official recognition will undoubtedly give further impetus to the already astounding growth rate of the Citizens Radio Service.

CB Licensing Procedures

Citizens Band Radio is the easiest U.S. communications service to enter. Anyone eighteen years of age or older may obtain a CB license. Only corporations owned by a foreign government are denied access to CB radio. Many individuals who possess no license utilize CB radios and do so legally. One license can cover a number of radios. If a person uses CB radio in his business, he may equip five, ten, or more service trucks as well as their base station under one license, and all these units can legally be used by the servicemen who drive the trucks but have no CB license of their own. Or a radio system can consist of a base station at home and a mobile unit in each of two or more family cars. The license will be issued to only one family member, but all family members can legally use the radios, including those below eighteen years of age.

Obtaining a Citizens Band Radio License is simple. No exam is required as it is for amateur or commercial class licenses. No Morse code must be learned. Three simple things are necessary for obtaining the CB license:

1. A copy of Volume VI, Part 95 of the Federal Communications Commission Rules and Regulations must be obtained, read, and kept in the CBer's possession.
2. FCC Form 505 must be completed and mailed to the FCC.
3. The Temporary FCC Permit Form 555-B must be filled out and kept with the gear until FCC Form 505 is received.

Volume VI, Part 95 of the FCC Rules and Regulations covers the Citizens Band, the Amateur Service, and Disaster Communications. The FCC requires that a CB radio operator own and be familiar with the Citizens Band Radio Service Rules and Regulations. Pamphlets which include the required information are available for a nominal cost at many CB suppliers. This is the least expensive way to obtain the rules as far as initial investment is concerned. Purchase of Volume VI from the Government Printing Office costs more, but included in the cost is an updating service which extends for an indefinite period of time. Each time the FCC changes a regulation relating to the Citizens Radio Service, it is sent to all Volume VI subscribers without further charge. Whereas the pamphlet will become obsolete and will have to be replaced, the Printing Office version will remain current. The address of the Government Printing Office is:

Superintendent of Documents
Government Printing Office
Washington, D.C. 20402

A copy of Part 95 of the FCC Rules and Regulations is supplied with each CB radio sold. This meets the CBer's immediate needs, but access to updated regulations is worth the fee charged by the Government Printing Office.

Figure 1-2 is a copy of Form 505. This new form is much simpler than its predecessor. The instructions are explicit and clearly stated. For legibility purposes, the form must be filled in with a typewriter, or capital letters must be printed in the boxes, leaving a blank box where a space would normally appear. The application must be signed and dated. Failure to sign and date the application will delay FCC action on it. Check with the FCC to see if a fee is required. If it is, payment must be enclosed in the form of a check or money order. Cash should not be sent. Order forms and subscription fees for Volume VI of the Rules and Regulations must not be included with the license form. These orders must be sent separately to the Government Printing Office at the address indicated above. The completed license application, signed and dated, must be mailed to the following organization:

The Federal Communications Commission
Gettysburg, Pa. 17326

Filling out the form is simple. It employs blocks for marking the appropriate response.

Item one is the applicant's name. Notice that the *first* name is to be written first.

Item two is the applicant's date of birth. The number of the month must be used.

Item three is used if the applicant is a business. If a business is not involved, this item is left blank.

Items four to seven provide for the street address, the city, the state, and the zip code of the applicant.

Items eight, nine, and ten describe the location of stations which have rural routes and P.O. boxes. The location should be given as specifically as possible (for example, corner of Church Street and Highway 183; two miles north of Junction 210 on RR 107).

Item eleven requires that the description which is most applicable to the applicant be checked.

Item twelve indicates the purpose of the application (that is, request for new license, renewal, increase in number of transmitters to be covered by the license). For renewals or requests for increase of the number of units to be authorized under the license, the current call sign of the station must be included.

Item thirteen identifies the request as being for Class C remote control operation or Class D voice communications service.

Item fourteen indicates the number of transmitters to be authorized by a license. If the requested number of units is sixteen or more, a statement justifying the need for that number of units must be attached to the application.

Item fifteen stipulates four things which the applicant must certify as true concerning himself:

1. The applicant certifies that he is not a representative of a foreign government.
2. He certifies that he owns or has placed an order for Volume VI Part 95 of the FCC Rules and Regulations.
3. He certifies that he will operate his Citizens Band Radio equipment according to the FCC Rules and Regulations.
4. He certifies that he has no legal claim against the United States Government that concerns the regulation of radio communication.

Item sixteen is for the applicant's signature.

Item seventeen is for dating the application at the time at which it is signed.

United States of America
Federal Communications Commission

Form Approved
GAO No. B-180227(R01 02)

FCC FORM 505

April 1976

APPLICATION FOR CLASS C OR D STATION
LICENSE IN THE CITIZENS RADIO SERVICE

INSTRUCTIONS

A. Print clearly in capital letters or use a typewriter. Put one letter or number per box. Skip a box where a space would normally appear.

B. Enclose appropriate fee with application. Make check or money order payable to Federal Communications Commission. DO NOT SEND CASH. No fee is required of governmental entities. For additional fee details see FCC Form 76-K, or Subpart G of Part 1 of the FCC Rules and Regulations, or you may call any FCC Field Office.

C. **Mail application to Federal Communications Commission, P.O. Box 1010, Gettysburg, Pa. 17325**

NOTICE TO INDIVIDUALS REQUIRED BY PRIVACY ACT CF 1974

Sections 301, 303 and 308 of the Communications Act of 1934 and any amendments thereto (licensing powers) authorize the FCC to request the information on this application. The purpose of the information is to determine your eligibility for a license. The information will be used by FCC staff to evaluate the application, to determine station location, to provide information for enforcement and rulemaking proceedings and to maintain a current inventory of licensees. No license can be granted unless all information requested is provided.

1. Complete **ONLY** if license is for an Individual or Individual Doing Business AS

FIRST NAME   INIT   LAST NAME

2. DATE OF BIRTH

MONTH   DAY   YEAR

3. Complete **ONLY** if license is for a business, an organization, or Individual Doing Business AS

NAME OF BUSINESS OR ORGANIZATION

4. Mailing Address

4A. NUMBER AND STREET

4B. CITY   4C. STATE   4D. ZIP CODE

(See reverse side of this form, for filling in Item 4C.)

5. If you gave a P.O. Box No., RFD No., or General Delivery in Item 4A, you must also answer items 5A, 5B, and 5C.

5A. NUMBER AND STREET WHERE YOU OR YOUR PRINCIPLE STATION CAN BE FOUND
(If your location can not be described by number and street, give other
description, such as on RT. 2, 3 mi., north of York.)

5B. CITY   **5C.** STATE

(See reverse side of this form for filling in Item 5C.)

6. Type of Applicant **(Check Only One Box)**

☐ Individual   ☐ Association   ☐ Corporation

☐ Business Partnership   ☐ Governmental Entity

☐ Sole Proprietor or Individual/Doing Business As

☐ Other (Specify) _______________

7. This application is for

☐ New License

☐ Renewal

☐ Increase in Number of Transmitters

IMPORTANT
Give Official FCC Call Sign

8. This application is for **(Check Only One Box)**

☐ Class C Station License
(NON-VOICE—REMOTE CONTROL OF MODELS)

☐ Class D Station License (VOICE)

9. Indicate number of transmitters applicant will operate during the five year
license period **(Check Only One Box)**

☐ 1 to 5   ☐ 6 to 15   ☐ 16 or more (Specify No.
and attach statement justifying need.)

10. CERTIFICATION I certify that:

• The applicant is not a foreign government or a representative thereof.

• The applicant has or has ordered a current copy of Part 95 of the Commission's
rules governing the Citizens Radio Service. See reverse side for ordering information.

• The applicant will operate his transmitter in full compliance with the applicable
law and current rules of the FCC and that his station will not be used for any purpose contrary to Federal, State, or local law or with greater power than authorized

• The applicant waives any claim against the regulatory power of the United States
relative to the use of a particular frequency or the use of the medium of transmission of radio waves because of any such previous use, whether licensed or unlicensed.

**THIS APPLICATION WILL NOT BE PROCESSED UNLESS
SIGNED AND DATED.**

**WILLFUL FALSE STATEMENTS MADE ON THIS FORM OR AT-
TACHMENTS ARE PUNISHABLE BY FINE AND IMPRISONMENT.
U.S. CODE, TITLE 18, SECTION 1001.**

11. SIGNATURE   **12.** DATE

Signature of: Individual applicant, partner, or authorized person on behalf of a
governmental entity, or an officer of a corporation or association

Fig. 1-2 FCC Form 505 license application.

These seventeen simple items constitute the entire application blank for a Citizens Band Radio License. The entries printed in boldface type should be given special notice. One of these warnings indicates that the operation of an unlicensed radio is illegal. A second boldfaced entry warns against willful false statements, stating that they can lead to fines and imprisonment.

The first of the warnings has been all but nullified by a recent FCC decision to permit a type of point-of-sale licensing. A person who purchases a CB radio can now obtain a Temporary Permit, Form 555-B, shown in Fig. 1-3. This form is completed when the unit is purchased and represents the CBer's authority to operate his CB rig for up to sixty days while his form 505 is being processed by the FCC.

Form 555-B is very simple. The front side of the sheet is divided into four parts. Part one indicates to the applicant when the Temporary Permit is required. Very simply, Form 555-B is required only during the period of time between mailing of Form 505 and its return.

Part two is the certification section. It requires that five things be certified in order for a person to be eligible for the Temporary Permit. Name, address, and signature are also requested in this section. Notice that the date on which Form 505 was mailed to the FCC is requested.

Part three assigns a temporary call sign to the station. The assignment method is ingenious. The temporary call sign is determined as indicated below:

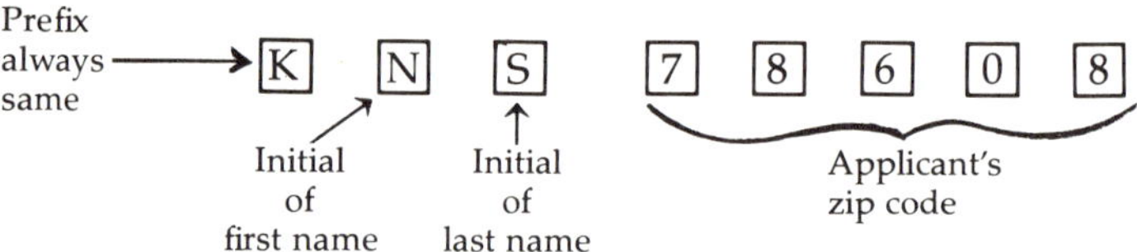

Part four outlines the limitations of this Temporary Permit. They are essentially the same as those of the Permanent License.

The reverse side of the sheet provides a very brief and basic listing of the requirements of the FCC concerning CB radio. While it is not complete, it does provide an understanding of the most salient portions of the Rules and Regulations that apply to CB operators. Two matters should be given special attention. Point two specifies that the Temporary Permit will authorize a maximum of five transmitters. Ten or more units may be requested on Form 505, but only five can be used prior to receipt of the Permanent License. Point four indicates that the operator must place a photocopy of Form 555-B at the fixed station location. Moreover, a card indicating the name, address, and temporary call sign of the applicant must be attached to each transmitter, mobile or fixed. Figure 1-4 shows

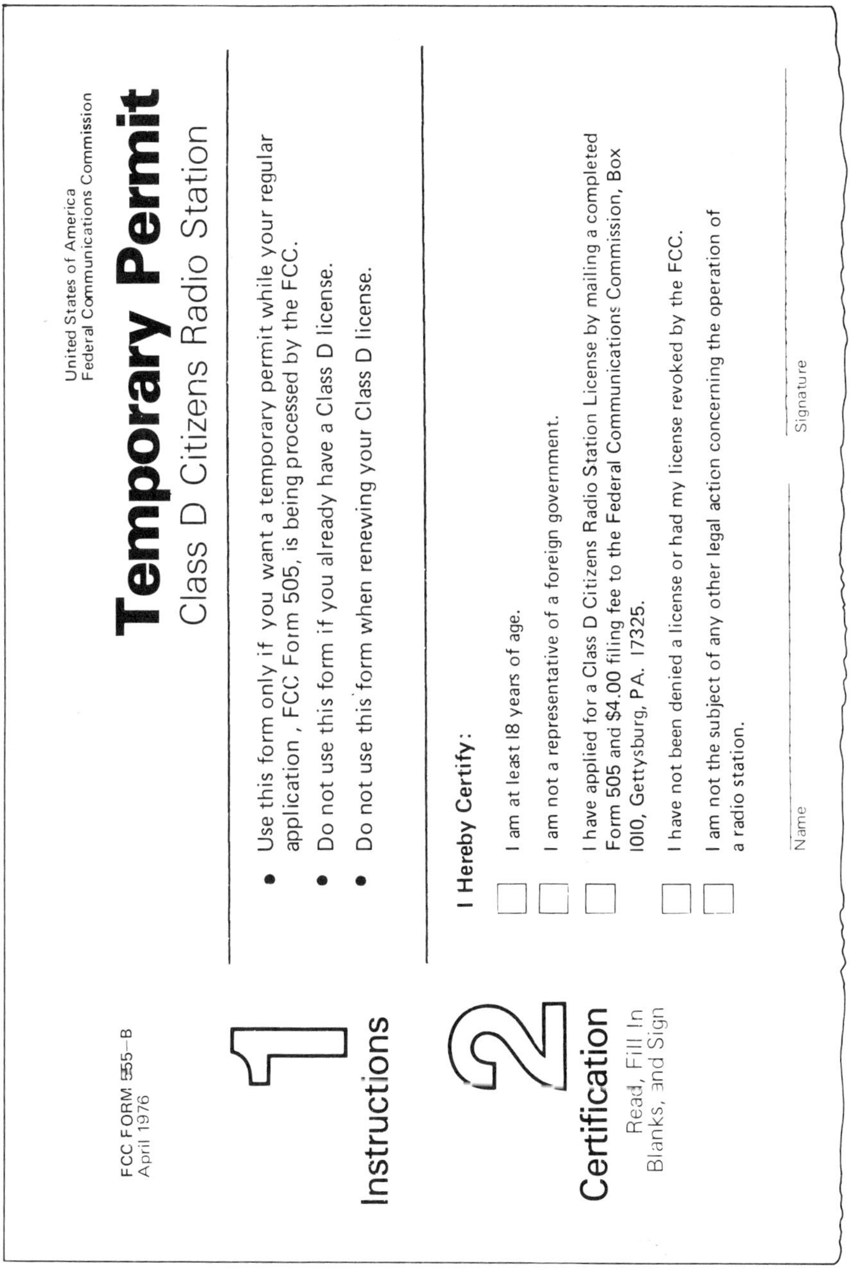

FCC FORM 555–B
April 1976

United States of America
Federal Communications Commission

Temporary Permit

Class D Citizens Radio Station

1

Instructions

- Use this form only if you want a temporary permit while your regular application, FCC Form 505, is being processed by the FCC.
- Do not use this form if you already have a Class D license.
- Do not use this form when renewing your Class D license.

2

Certification

Read, Fill In
Blanks, and Sign

I Hereby Certify:

☐ I am at least 18 years of age.

☐ I am not a representative of a foreign government.

☐ I have applied for a Class D Citizens Radio Station License by mailing a completed Form 505 and $4.00 filing fee to the Federal Communications Commission, Box 1010, Gettysburg, PA. 17325.

☐ I have not been denied a license or had my license revoked by the FCC.

☐ I am not the subject of any other legal action concerning the operation of a radio station.

_______________________   _______________________
Name   Signature

Fig. 1-3 FCC Form 555-B temporary permit.

3 Temporary Call Sign

Address

____

If you cannot certify to the above, you are not eligible for a temporary permit.

Willful false statements void this permit and are punishable by fine and/or imprisonment.

Date Form 505 mailed to FCC

- **Complete the blocks as indicated.**

  Use this temporary call sign until given a call sign by the Federal Communications Commission.

K □ □ □□□□□

Initial of Applicant's First Name

Initial of Applicant's Last Name

Applicant s Zip Code

4 Limitations

Your authority under this permit is subject to all applicable laws, treaties and regulations and is subject to the right of use or control by the Government of the United States.

This permit is valid for 60 days from the date the Form 505 is mailed to the FCC.

You must have a temporary permit or a license from the FCC to operate your Citizens Band radio transmitter.

Do Not Mail this form, it is your Temporary Permit.

See the reverse side of this form for a summary of operating instructions.

Fig. 1-3 FCC Form 555-B (cont'd).

1. ________________   UNITED STATES OF AMERICA   FCC FORM 452-C
   (STATION CALL SIGN)   FEDERAL COMMUNICATIONS COMMISSION   (Revised)

TRANSMITTER IDENTIFICATION CARD

THIS CARD ATTESTS THAT AUTHORIZATION HAS BEEN RECEIVED FROM THE F.C.C. FOR
INSTALLATION AND/OR OPERATION OF THE-RADIO TRANSMITTER TO WHICH ATTACHED.

2. NAME OF PERMITTEE OR LICENSEE ________________________________

3. LOCATION(S) OF TRANSMITTER RECORDS ________________________________

4. TRANSMITTER OPERATING FREQUENCIES ____________________CLASS D CITIZENS BAND

5. SIGNATURE ________________________________
              (PERMITTEE, LICENSEE, OR RESPONSIBLE OFFICIAL THEREOF)

CURRENT FCC AUTHORIZATION FOR THIS TRANSMITTER EXPIRES ________________________

EBP-189   104067

Fig. 1-4 FCC Form 452-C identification card.

FCC Form 452-C, which may be attached to the radio to provide the required identification during the Temporary License period. This form will continue to be used when the Permanent License is received.

Limitations and Potentials of CB Radio

Class D Citizens Band Radio is a lot of things to a lot of people, but it is not all things to all people. For one thing, CB radio is limited in range. The FCC has restricted the power of CB radios, and the 11 meter band (see Fig. 1-5) does not normally permit long distance communication. The power limitation is the result of international treaty and FCC decisions. The band conditions are the result of the propagation characteristic of the band.

The upper atmosphere of the earth is less protected from sun generated rays than the lower atmosphere. Several layers of gases in the upper atmosphere form what is called the *ionosphere*. The gases in the ionosphere react to bombardment of their molecules by the sun's radiation by forming charged particles called *ions*. Different layers of the ionosphere react differently and to varying extents. The importance of the ionosphere for CB communications is that the ionized layers reflect radio waves which strike them. The degree of reflection is determined by several factors. Ionization of one layer of the upper atmosphere (the E layer, see Fig. 1-6) increases in direct proportion to the amount of radiation the gases receive from the sun. This means that ionization in the E layer is greater in the summer than in the winter and builds to a peak at midafternoon. Another factor in upper atmosphere ionization is the solar phenomena referred to as *sun spots*. Sun spots increase and decrease in number and intensity. These variations occur in overlapping cycles. Smaller cycles increase to a maximum and then decrease. At the same time, a larger cycle is active so that the peak of each smaller cycle is greater than the one before it until a maximum is reached. This larger cycle

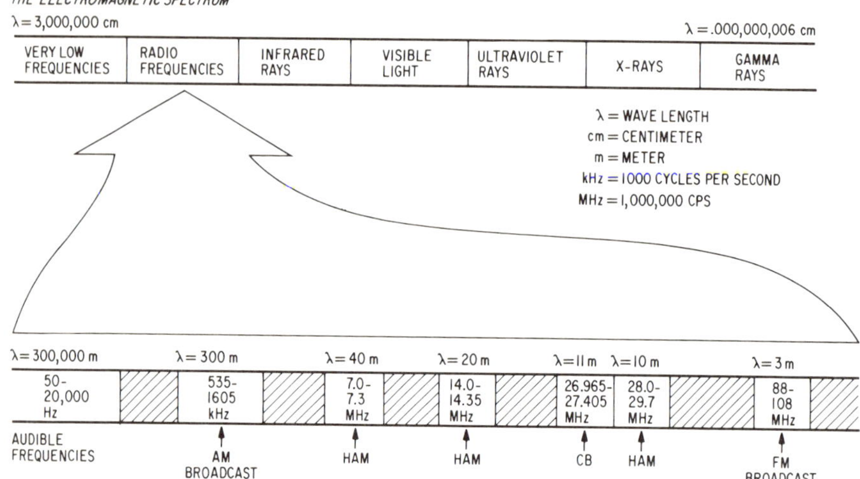

Fig. 1-5 *The radio frequency spectrum:* (a) Radio waves lie in a spectrum of frequencies between subaudible frequencies and light. (b) Radio waves increase and decrease in strength in a cyclic pattern. Since the waves travel at the speed of light, a set distance is traveled during one cycle by each specific frequency. CB radio waves travel a distance of 11 meters as one cycle is completed.

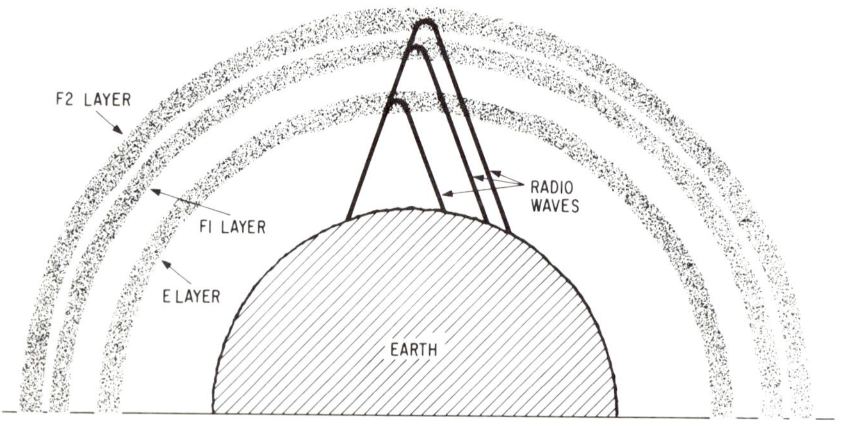

Fig. 1-6 *Layers of the Ionosphere:* The gases of the upper atmosphere ionize in different layers. The E layer and the F layers have the greatest effect on CB radio signals. The E layer responds to ultraviolet radiation, and the F layers vary according to sunspot activity.

requires eleven years to complete itself. The result of all this cyclic change is that the way the F layers of the ionosphere react to radio signals is constantly changing. Reflection of signals by the F layers is greatest at the peak of the sun spot cycle and during winter days. Winter cooling contracts the upper atmosphere, and this contraction causes an increase in signal reflection.

Another factor which determines how the radio wave will be reflected by the ionosphere is the frequency of the radio wave. Lower frequencies tend to be absorbed by the ionosphere. Very high frequencies punch through it and are lost in space. CB frequencies either pass through the ionosphere when it is thin or are reflected when the ionization increases.

One other factor which determines whether a radio wave is reflected or not is the angle at which it strikes the ionosphere. As the angle of incidence between the radio signal and ionosphere becomes steeper (that is, closer to 90 degrees), the more likely the wave is to pass through and not be reflected.

For the intended local communication of CB radio, the ionosphere has little effect. The most it can do is to cause signals from great distances to be reflected into the local area, thus interfering with local communications. This reflection type of propagation is called *skip.* It is possible for stations thousands of miles away to skip into the local area at a greater signal strength than that of local stations. Communication could therefore take place between stations thousands of miles apart. Unfortunately, this "skip" communication is prohibited. CB communications are limited by FCC regulations to 150 miles or less. Because CB radio depends primarily on line-of-sight communication, the actual range is limited to a radius of 25 miles or less under normal circumstances. This range is usually adequate.

Another limitation of CB radio is the crowding of the CB band. This congestion has been partly alleviated by the expansion of the band. Prior to January, 1977, only 23 channels were available to CB enthusiasts. At that time 17 additional channels were made available so that there are now 40 channels for public use (see Table 1-1). Even forty channels are not expected to satisfy the increasing demands on CB radio. Other possibilities for frequency expansion are therefore under consideration by the FCC. One possibility is the relocation of the Citizens Radio Service to a band of frequencies in the 200 or 900 MHz range. A large number of channels would result from this allocation, but new types of equipment would be required.

CB radio is destined to become a larger service. Too many people are interested in it for its potential to be left cramped and unexploited. The CB fraternity will not only grow but increase in power.

TABLE 1-1   Frequencies Available to the Citizens Radio Service

Channel No.	MHz	Channel No.	MHz
1	26.965	21	27.215
2	26.975	22	27.225
3	26.985	24 ⎫	27.235
4	27.005	25 ⎬ *	27.245
5	27.015	23 ⎭	27.255
6	27.025	26	27.265
7	27.035	27	27.275
8	27.055	28	27.285
9	27.065	29	27.295
10	27.075	30	27.305
11	27.085	31	27.315
12	27.105	32	27.325
13	27.115	33	27.335
14	27.125	34	27.345
15	27.135	35	27.355
16	27.155	36	27.365
17	27.165	37	27.375
18	27.175	38	27.385
19	27.185	39	27.395
20	27.205	40	27.405

*Note: The FCC does not refer to the frequencies by channel number. The channel numbers shown above have been agreed upon by the CB radio manufacturers. With the reallocation of January 1, 1977, frequencies 27.235 and 27.245 were made available for CB use. These frequencies were given the designations Channels 24 and 25 respectively, even though Channel 23 is higher in frequency, so as to maintain compatibility with existing 23 channel units.

CHAPTER 2

Operating on the Citizens Band Fixed Service

Foundations of CB Communications

Communication, needless to say, requires more than one person. One person speaks and one or more listens. Normally, an interchange is involved such that after the first person finishes speaking, he becomes a listener while another person speaks. Radio is simply an electronic extension of face-to-face conversation. Using radio, the speaker transmits and the listeners receive. Then another transmits while the first receives. All that radio leaves out of the communication is the mustard on the young man's chin and the scent of perfume which surrounds the pretty young miss.

There are two basic types of Citizens Band radio stations, the base station and the mobile station. A third class of station, the portable station, is a special form of mobile station providing mobility without the vehicle. Communication can take place between any combination of station types. Base stations may communicate with one another. A base station and a mobile station may have interchanges. Mobile stations may carry on conversations with other mobile stations, and portable stations may communicate with other portable stations or with mobile or base stations. Each type of interchange has its purposes and its limitations.

Range differs markedly between the various communications systems. Table 2-1 indicates the relative distances to be expected between the various types of stations. As can be seen, base-to-base communications have the greatest range, but they are also the least flexible. A base station is fixed in position. For communication to take place between people using base stations, the individuals must be at the right place at the right time. Obviously, if they are not at their base locations at the same time, they cannot communicate.

Base-to-mobile communication is more flexible. One individual can be in motion or in any of a myriad of locations and still be able to communicate with the base station. This flexibility is gained at the ex-

TABLE 2-1   Range Vs Flexibility of Various System Combinations

System Combination	Range	Flexibility
Base-Base	Greatest	Limited
Base-Mobile	Moderate	Moderate
Mobile-Mobile	Limited	Very flexible
Portable-Base	Limited	Moderate
Portable-Mobile	Very limited	Very flexible
Portable-Portable	Most limited	Most flexible

pense of range, however. The mobile station uses an antenna with less gain and located at a very low height (the height of a car). These antenna restrictions reduce the effective range of the system.

Mobile-to-mobile communication represents the maximum of flexibility. All parties can now be in motion and in constantly changing locations and yet be able to communicate. Again, the flexibility is won at the sacrifice of range. Mobile-to-mobile communication is undependable outside of a very limited range. If a base station is included with the mobile units, greater range can be obtained by relaying messages through the base. Portable units magnify the range versus flexibility considerations.

Characteristics of CB Base Stations

The base station is included under the description "fixed service." A base station is usually the "home" unit of a system. Citizens Band licenses cover a system of radios and not just a single unit. Therefore, if a person installs a unit at his house and one in his car, only one license is required and only one call will be issued. When communications between the car and home take place, the units are identified by referring to the fixed station as "Base" and the mobile station as "Unit One." If a second base station is installed at an office, even though it is a fixed station, it would be referred to as "Unit Two."

A base station in its simplest form consists of the radio equipment and the antenna. With the rise of CB radio as a hobby, however, the base station has begun to include more items. Many accessories are now considered a vital part of the base station. Desk top microphones are in wide use, as are station clocks, area maps, phone patches, and special speakers. Indeed, the avid CBer may designate a desk or a whole room as his station. He may invest in multiple antennas, scanning receivers, remote reading weather instruments, and any number of other exotic accouterments. The base station CB operator often visualizes himself as being in the business of providing a service to his fellow CBers. He will not stint on time or expense in fulfilling the obligations of his post.

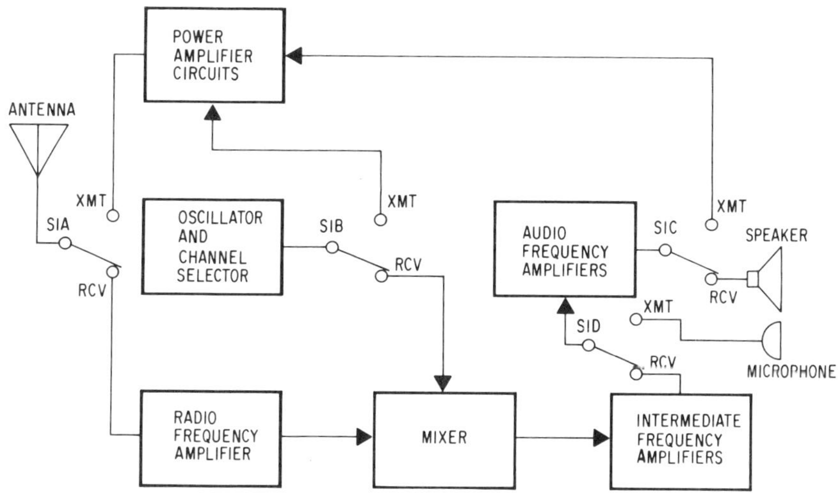

Fig. 2-1 *Block diagram of a typical transceiver:* S-1 is switched to transmit when the microphone switch is pressed.

Most Citizens Banders use transceivers, primarily because of their cost and operating convenience. A transceiver, as indicated by its name, combines a transmitter and receiver into one unit. A reduction in manufacturing cost is realized by constructing the unit in this way. Figure 2-1 illustrates how the components of a transceiver are interrelated. Some of the components are used in both the transmitter and receiver sections of the unit. For example, the audio components are used as a microphone preamplifier during transmit and are used in the speaker amplifier circuit during receive. Oscillator circuits, crystals, and other components also serve dual functions in transceivers. This reduction in the required number of components is reflected in decreased size and cost.

A transceiver is an integrated unit. Changing from transmit to receive is accomplished automatically by circuits built into the radio. If a separate receiver and transmitter were used, the transmitter would have to be turned off and the receiver turned on manually, a less convenient arrangement. Automatic switching devices are available, but they entail extra cost and increased complication. The problem of tuning the receiver to the same frequency as the transmitter, for example, represents a marked decrease in operational simplicity. In view of the added cost and decreased convenience, one may wonder why separate units, such as those shown in Fig. 2-2, would be wanted at all. There are reasons. A person may desire to be able to tune the receiver across the CB channels without moving the transmit frequency. Also, separate receivers are normally of higher quality than the receiver section of a transceiver. A

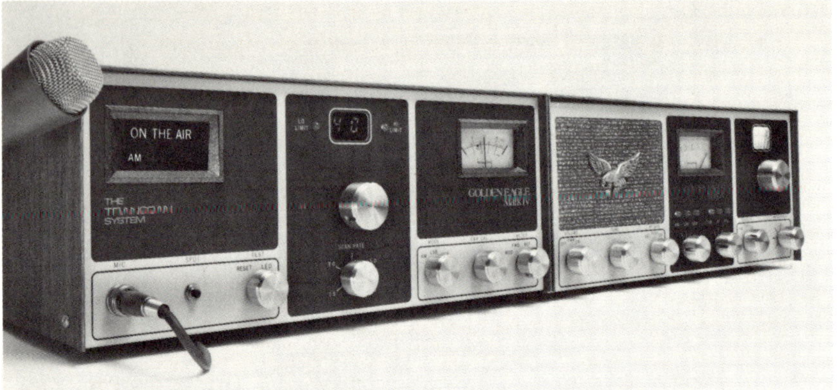

Fig. 2-2 Separate units for base station use. (*Courtesy*, Browning Laboratories, Inc.)

transceiver may be combined with a separate receiver so that a conversation can be in progress on one channel at the same time that emergency channel 9 or some other channel is being monitored. The choice of base station equipment depends to a large degree on the use to which a person puts his CB radio. As indicated, the majority of CBers use transceivers.

Base station transceivers have some differences from mobile transceivers. Most of these differences arise from the stationary nature of the fixed service. The base station transceiver often possesses more dials and knobs than mobile units. Understandably, a person has greater opportunity to tune transceiver controls and observe transceiver dials when he is relaxing in a deeply cushioned desk chair than when he is weaving in and out of five o'clock traffic.

The base station transceiver is usually larger than mobile units. One of the cigarette-pack-sized mobile transceivers looks lonesome on a large desk top. Often the base station unit is built for attractiveness rather than ruggedness. It will suffer less jolts and bumps at home than under the dash of a car.

It should be emphasized that regardless of how large or how pretty the base station transceiver may be, it cannot legally transmit any more power than the mobile unit does. It may have extra features that make operating easier and more fun, but it will have no more talk power than a good mobile transceiver. Indeed, many mobile transceivers make very adequate base station rigs when operated with an appropriate power supply.

Base Station Installation

Installation of Citizens Band equipment at a base location can be a very simple process or a very difficult ordeal. A lot depends on the type of equipment selected and the nature of the building in which it is to be installed. A base station transceiver is usually ready to operate directly out of the box. All that is needed is to connect a microphone and an antenna, plug the transceiver in, turn it on, and start talking. There are times, however, when these simple steps can encounter complications.

If the unit is to be placed in the CBer's home, installation procedures should follow the path of least difficulty. However, this does not mean that all problems are automatically eliminated. CBers have adopted a term coined by Amateur Radio operators—"XYL." The term "YL" in radio shorthand means an unmarried young woman. An "XYL" is a wife, the one who is most likely to fuss when the big black cable snakes through the partially opened window (warping the screen out of shape), across the bed, and in front of the chest-of-drawers before it gets to the CB radio. Certainly, this situation cannot be tolerated. How does the CBer get his antenna lead to his radio? If the radio is located next to the outside wall, a simple "feed through" bushing can be used. This device is nothing more than a hollow tube with special caps on each end which will seal off against the outside of the wall to stop water and insects from entering the house while permitting entry of the antenna transmission line. The device is installed by drilling a hole through the wall and then inserting the bushing. If the bushing is too long, it can be cut to fit. Caulking around the outside bushing cap will insure against leaks. If the house has a brick exterior, the installation is more difficult but still possible. It is necessary to use a masonry bit of the appropriate size to drill through the brick. Application of too much pressure may result in destruction of the bit or drill motor. It is easier to drill through the mortar joint, but it is also more difficult to obtain a weather tight seal on this uneven surface. A wood bit should be used after a hole is made through the masonry veneer. If the bit will not reach all the way through the wall, a spike can be placed in the hole and used to puncture the inside wall. A hole can now be drilled where the spike came out. Location of electrical wires must be determined prior to drilling in the wall. It is good practice to wear insulated gloves and to turn off the electricity when it is not needed. Figure 2-3 shows a typical installation of a feed-through bushing.

If the CB station is located against an inside wall, a different method of installation must be used. The antenna wire can usually be routed into the attic through the eave of the house. A hole must be cut in the inside wall, as shown in Fig. 2-4, and a hole drilled in the top plate of the wall. The cable can now be snaked down the wall until it can be pulled out of the hole cut in the wall. A self-clamping electrical box, as shown, can be installed in the hole. The cable must be inserted into the box before inserting the box into the hole. After the box is installed, a cover plate which has a cable connector mounted to it is placed on the front of the box. A short cable can now be used to connect to the CB set. It is often

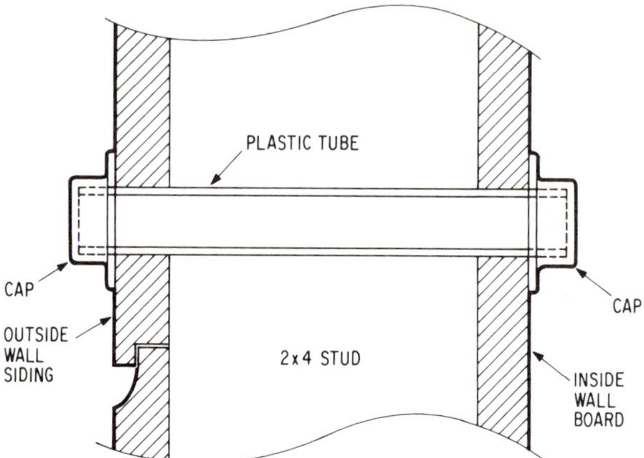

Fig. 2-3 Feed-through bushing installation.

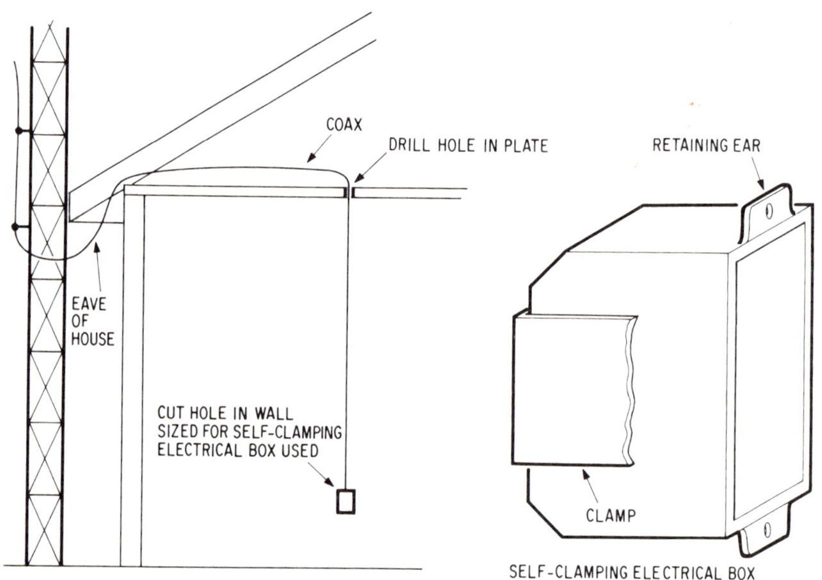

Fig. 2-4 Antenna cable installation for an inside wall.

necessary to run a stiff wire down the wall first and then use it to pull the coax into the box.

If the CB radio is located near a window, a simple cable entry device as shown in Fig. 2-5 can be easily constructed. A board long enough to reach from one side of the window to the other is cut in two, and thick rubber foam weatherstripping is attached to the two surfaces and to the ends. The sash is raised and the board is inserted at the bottom of the

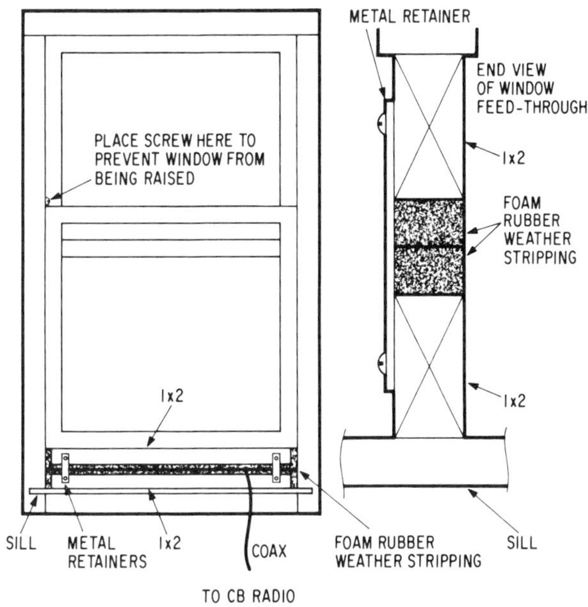

Fig. 2-5 Window feed-through device for antenna cable.

window. Antenna cables can now pass through the board between the upper and lower rubber foam. If metal retainers are included, the rubber foam can be compressed against the cable to seal out the weather. The screen must be modified to accommodate the cable. A similar feed-through arrangement can be used on the screen. If the window is seldom opened, the screen can be removed. Locking the window requires that a metal screw be placed above the sash to prevent it from being raised. It goes without question that any of these procedures undertaken in rental property must be first discussed with the owner or manager.

Grounding is important for base stations. Grounding of the transceiver not only can reduce the pick-up of noise, it can also prevent shock and reduce danger from lightning. A good earth ground will often improve the operation of the station. A ground is usually obtained in one of two ways. If the plumbing of the house uses metal pipe, a ground clamp can be attached to a water pipe at the point where the pipe enters the ground. If no water pipe is available, a ground rod can be driven into the ground. A No. 12 or larger copper wire should connect to the transceiver and to the antenna lightning arrestor. If a tower or mast is used, it, too, should be grounded. Figure 2-6 shows a typical grounding system.

Additional electrical plugs near the CB station are worthwhile investments. Overloaded and distant outlets can reduce operating efficiency and create fire hazards.

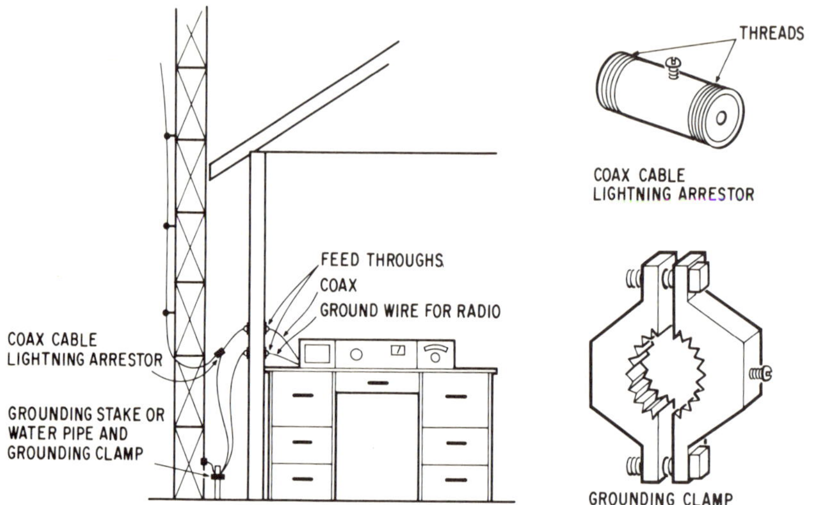

Fig. 2-6 Base station grounding system.

As indicated earlier, a base station is a personal thing. What a person includes in his station and how he sets it up involve individual choices.

Operating a CB Base Station

For many people, the hardest word ever spoken is the first word transmitted over Citizens Band radio. A large measure of the hesitancy is due to a lack of acquaintance with CB communications. The beginning CBer just does not know what to say. Fortunately, this fear is quickly dispelled once the person begins to operate his radio.

In actuality, CB communications have developed an informality under which almost anything goes *except* formality. This informality is one of the main reasons that CB radio has become so popular. After saying this, however, it must also be emphasized that some formality is desirable if communication rather than confusion is to result.

Contacting another station can follow one of several routes depending on the circumstances. If the base station desires to contact a mobile unit which is part of the same station, the call could follow this format:

"This is KXX 1234, Base, calling Unit One. Over." The reply could be:

"KXX 1234, Base, this is Unit One. Over."

Notice that the mobile unit is referred to simply as Unit One and not by the call sign. This is acceptable since the stations are both authorized under the same license and call sign. It is necessary to include one's own call sign in the transmission. The FCC requires the call sign to be given at the beginning and conclusion of each transmission. When a series of short transmissions take place, the call sign can be given only at the beginning and conclusion of the series or at intervals of no more than five minutes.

The contact given above was very formal. The more usual CB communication relies on "handles" or nicknames and less stylized techniques. The CBer must remember that FCC rules and regulations must be followed, informal or not. The following interchange is an example of a less formal contact initiated by the mobile unit:

"KXX 1234, Daddy Bear, this is Mamma Bear, rolling on in Mobile One; do you hear me? Over."

"I hear you, Mobile Mamma One; this is Daddy Bear, KXX 1234 Base. Come on."

Contacts between stations authorized by different licenses follow similar forms. In most cases, both call signs will be given. The FCC has made it legal for a station to communicate with another station without being responsible for giving the other station call. The CBer is responsible for transmission of his own call, however. Both of the following communications are legal:

"KXX 1234, Daddy Bear, this is KZZ 4321, Goldilocks. Mercy sakes, do you hear me? Come on."

"Mercy sakes, Goldilocks, you ole pea picker; this is Daddy Bear, KXX 1234. I sure do hear you. That's a Forty Roger; back to you."

There is one other major contact method which the CBer should be familiar with—the break. If a person is interested in establishing a contact but has no particular station in mind, he may choose to break into a conversation that is in progress between other stations. This procedure is understood by CBers, and the other stations will pause in their own conversation to admit the newcomer when possible. An example of establishing a contact by breaking follows:

". . . and I want to tell you, Daddy Bear, that was the biggest fish I ever saw. Come on."

"Break Nineteen . . ."

"OK, Catfish, we know that fish was big, sure do. Say we have a breaker that wants to get in here. Stand by Catfish while I pick her up. This is Daddy Bear, KXX 1234, talking with Catfish, KZX 1432. Come on, Breaker."

"KXX 1234 and KZX 1432, this is KZZ 4321, Goldilocks. Thanks a lot Daddy Bear for letting me in here. I heard Catfish talking about his fish and. . . ."

The examples given are representative of the exchanges the CBer will hear on the band. Different operators may have their own styles, but the basic message will remain the same.

Once the CBer learns what to say on the Citizens Band, he needs to learn how to adjust his radio so that he can say it. Although there are so many different radios available that it would be impossible to give instructions for each one, enough similarities exist among the various units for some general instructions to be given.

When operating the CB unit for the first time, it is best to turn off all accessories such as the *noise blanker* and the *automatic noise limiter* if the unit has these. The next step is to choose the channel on which to operate. The best source of information on what channel should be used in any particular area for the type of communications the CBer plans is another local CB operator. If access to a local CBer is difficult, listen to the various channels and pick the one that seems to be the most appropriate. The channel is chosen by turning the *channel selector* until the desired channel number is at the *operate location.* In most transceivers, both transmitter and receiver will be set for operation on the selected channel.

Two or three controls determine what is heard from the speaker. All units have a volume control of some kind. It may be called *Volume, AF Level,* or something similar. This control determines how loud the sound from the speaker will be. It adjusts the gain of the audio or sound amplification circuits. It should be set to a comfortable listening level.

A *squelch* control is provided on many units. This control is adjusted to cut the sound off from the speaker when the received signal strength is not greater than that of the background noise. The best way to set the squelch control is to adjust it when no signal is present on the channel. The control should be adjusted until the noise is cut off and the speaker is quiet. When a station transmits on the channel, the higher received signal level will "break squelch," and the transmission will be heard. When the transmission ceases, the noise level will not break squelch and the speaker will be silenced again. It is possible to adjust the squelch to a point at which no signals or only very strong signals will be heard. It is a good practice to check the squelch each time operation is begun. Also, if it is desirable to receive weak signals, it may be necessary to turn the squelch to minimum.

Some units provide an *RF level control.* This control operates in similar ways to the AF level control except that it controls the early stages of the radio. The RF control is sometimes referred to as the *sensitivity control.* It is set to provide the desired signal level with the least noise. The stronger the desired signal is, the lower the RF level control can be set. The weaker the signal is, the higher the control must be set and the louder will be the background noise and the other signals. There is some interaction between the RF level control and the volume control in that both increase the speaker volume. The best way to set the two is to advance the volume control to a proper setting and then to reduce the RF level control to reduce the noise. The volume control will have to be advanced as the other is reduced. Between contacts, it is best to advance the RF level control to maximum and to reduce the volume control if necessary. This insures that even weak signals will be heard if they break squelch.

If the radio employs accessories such as a *noise blanker* (NB), *automatic noise limiter* (ANL), or an *automatic gain control* (AGC), switches may be provided to turn these on and off as desired. It is usually possible to leave all of these circuits on during normal operation. However, there may be special situations when operation will be improved by turning them off. The noise blanker and noise limiter can be turned on and off while listening to the speaker. They should be set at the position which gives the best sound. The AGC reduces the gain of the radio depending on the strength of the received signal (it is actually an automatic RF level control). This action is usually desirable. If the CBer is operating when lightning is present or when a local station is strong and the desired station is weak, he may desire to set the level so as to permit reception of the desired signal even though the distracting signal is louder than usual. In this case, it may be advantageous to turn off the AGC.

Single side band (SSB) transceivers have some controls not found on AM units. Since the SSB units can use either of two sidebands for each AM CB channel, a switch is provided to permit the desired mode of operation to be chosen. If AM is chosen, the unit operates like any other AM CB transceiver. If upper side band operation is desired, the unit transmits and receives on the upper side of the AM channel indicated by the channel selector. Operation on the lower side band is another option.

Some units provide a control to adjust the frequency of the receiver by small amounts. The control may be called a *clarifier, delta tune, receiver tune*, or something similar. It is used to tune the receiver to obtain the best reception. Such tuning is necessary if the two transceivers are not exactly on frequency. SSB signals begin to sound garbled (like a duck) and can become unintelligible when the frequency is not correct. More stringent frequency requirements have been proposed for future SSB units. Until then, the receiver tuning controls will be needed.

Most CB transceivers employ an *"S" meter*. The "S" stands for "strength." The meter provides a reading of the signal strength of the received signal. Signal strength is a relative measure which is described by a scale of one to nine. A very strong signal would indicate nine on the meter, whereas signals of less strength would show some lesser number. On occasion, the CBer may hear a report given in the form of "five by nine." The "five" is a description of readability. Distortion, overmodulation, and various types of equipment malfunction can cause the audio signal to be hard to understand. Readability, rated on a scale of one to five, is a subjective rating of the operator (see Table 2-2).

On some CB transceivers, the "S" meter serves a dual function. It measures received signal strength in the receive mode and indicates

TABLE 2-2   Signal Ratings

Readability

1 Totally unreadable
2 Barely readable; some words understandable
3 Difficult to understand
4 Understandable with little difficulty
5 Perfectly readable

Strength

1 Extremely weak signals; almost undetectable
2 Very weak signals
3 Weak signals
4 Fair signals
5 Fairly good signals
6 Good signals
7 Fairly strong signals
8 Strong signals
9 Very strong signals

modulation in the transmit mode. The meter is useful for both functions. On some portable units the meter is used to check the battery in addition to its other functions.

Selecting a CB Radio

As the number of CB operators has reached staggering proportions, the number of CB equipment manufacturers has increased in a similar fashion. The result is that there is a bewildering array of CB radios available. The beginning CBer may find himself in a dilemma concerning which unit to purchase. The final selection is usually made as a result of compromises between several criteria: proposed use, technical quality, provisions desired, budget, and personal preference. Assistance can be provided in evaluating the available equipment in terms of the first four criteria. The final criterion must be evaluated by the CBer alone.

Each CB radio has its own personality (see Fig. 2-7). Teaberry's Model "T" conveys a bit of nostalgia. E. F. Johnson's Messenger 4250 packages practicality in a space age design. Browning's Baron says elegance. SBE's Aspen is "sweet simplicity." The choice has to be made according to the proposed use. If the CBer desires a rugged mobile unit, he will select from one set of radios. If he wants a top-notch base station that he can brag about, he will choose from a different selection. The CBer must decide what type of unit he really wants and what he wants to do with it. One thing is certain. There is a CB radio which has a personality to match his own.

The options and accessories provided by CB radios vary greatly. The purchaser must choose between AM and SSB operation. He must decide between base station console or mobile transceiver. Should the unit

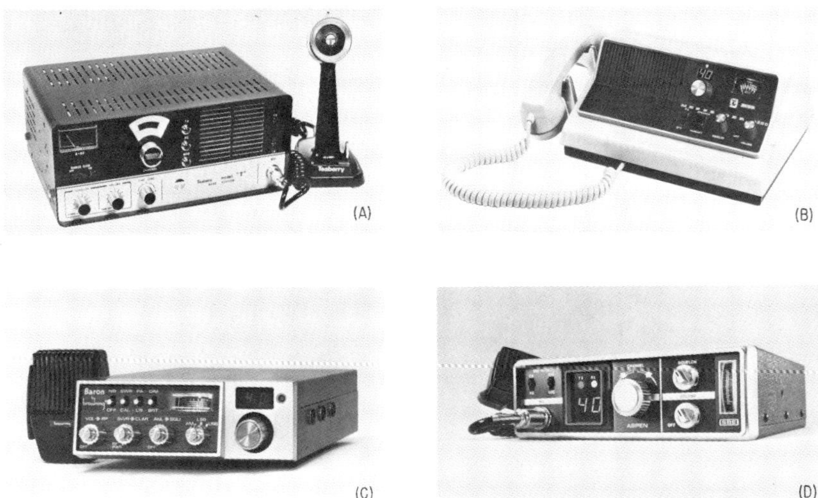

Fig. 2-7 *The personalities of CB radio:* (A) Nostalgia, (B) practicality, (C) elegance, (D) sweet simplicity. (A) *Courtesy,* Teaberry Electronic Corp.; (B) *Courtesy,* E. F. Johnson Co.; (C) *Courtesy,* Browning Laboratories, Inc.; (D) *Courtesy,* SBE Linear Systems, Inc.

possess noise reduction devices, delta tuning, signal strength metering, or a power microphone? Should the transceiver use traditional crystal circuitry, or should it afford the more modern synthesized design? The CBer must investigate the nature of each option and then evaluate his own needs and budget to decide if he wants to take advantage of it.

Budget is a universal factor in CB radio selection. Each person must decide what his own level of spending will be and whether he will pay cash or finance. Some will buy less than they want while others will buy more than they need. If there is a word of guidance to be given, it is to buy exactly what one wants if that is at all possible. Satisfaction is much more likely if one gets what he really desires than if he settles for something he feels is inferior. In a competitive market such as CB radio has become, the buyer can often benefit from sales and special promotions. It is wise to shop widely prior to purchase.

After the CBer narrows his choice to a few units which possess the features he wants, there are several possibilities for help in making the final decision. Salesmen at CB stores are usually good sources of information. If some of the units being considered include brands a particular shop sells, the salesman can probably give you an honest comparison of the various models. He can suggest which one is most popular and may even be able to provide names of customers who have purchased the different sets. Your final decision will be more valid if based on discussions with salesmen from a number of stores.

People who own the sets you are interested in are good sources of information. One must use caution in these discussions, however. People who own a particular CB set tend to react in one of two extremes. Either they think their set is the best, or they are very unhappy with it. The prospective buyer should get specific reasons for these likes and dislikes before he makes his own decision.

TABLE 2-3   SBE Cortez 40 Specifications*

Receiver Section

Circuit type	Double conversion superheterodyne with ceramic bandpass filter
Frequency	40 channels in the 27 MHz Citizens Band
Sensitivity	Less than 1 μV for 10 dB (S+N)/N
Intermediate frequency circuitry	Double conversion
Audio output	3 W at less than 10% THD
Receiver current drain	0.7 A squelched; 1.2 A full audio

Transmitter Section

Frequency	40 channel, phase-locked loop (PLL) digital frequency synthesis
RF power output	4 W maximum
Adjacent channel rejection	−60 dB
Modulation	AM
Transmitter current drain	1.9 A full AM modulation
Power source	13.8 V dc

*Courtesy, SBE, Inc.

TABLE 2-4   Hy-Gain V/Model 2705 Specifications*

Circuitry complement	35 transistors, 40 diodes, 1 IC, 4 FET's
Sensitivity	AM: 1 V for 10 dB (S+N)/N
	SSB: 0.25 V for 10 dB (S+N)]N
Selectivity	AM: 6 kHz at 6 dB down
	SSB: 2.4 kHz at 6 dB down
RF output power	SSB: 12 W PEP
	AM: 4 W (output)
Fine tune range (Clarifier)	±600 Hz
Squelch range	AM: 1 V to 10 mV
	SSB: 0.7 V to 20 V
Audio output power	3.0 W
Channels	40 AM or 80 SSB; phase-locked loop circuitry
Meter	Signal strength/RF
Frequency response	400 Hz to 2.6 kHz (+3dB, −10dB)
Carrier suppression	40 dB down
Harmonic suppression	60 dB down
Power supply	13.8 V dc, positive or negative grounding
Size	2¾'' H x 8½'' W x 11½'' D
Shipping weight	8.8 lb

*Courtesy, Hy-Gain Electronics Corp.

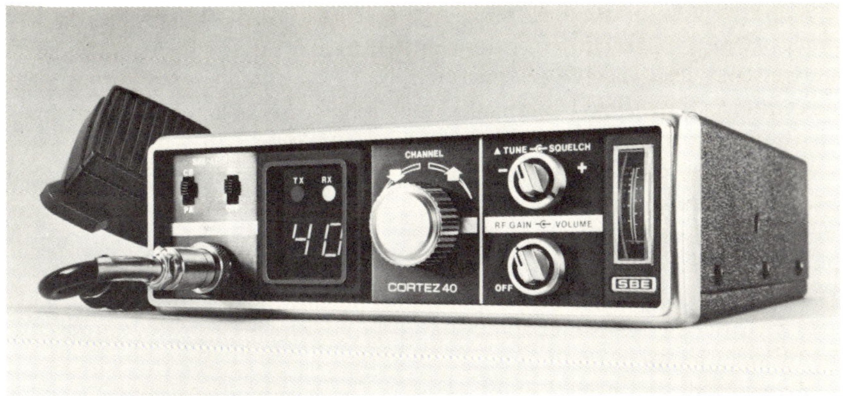

Fig. 2-8 The SBE Cortez 40 CB Transceiver (*Courtesy*, SBE, Inc.)

One other source of information for equipment selection is to be found in the equipment specifications. To make proper use of them, one must understand what they mean and which of them are most important. Two sets of specifications are given here to show how they may be evaluated. Specifications for the SBE Cortez 40 (see Table 2-3) are typical of those prepared for AM transceivers. Specifications for the Hy-Gain V/Model 2705 (see Table 2-4) are representative of SSB units. Both these units have commendable specifications.

The SBE Cortez 40 shown in Fig. 2-8 utilizes a double conversion superheterodyne receiver circuit with ceramic filter. To understand the significance of this description, one needs to gain some familiarity with the terms. Citizens Band radios transmit information as a radio frequency signal in the 27 megahertz range (11 meters). This is a comparatively high frequency and is difficult to amplify. To make the required amplification easier, the 27-megahertz signal is converted to a lower frequency where amplification is simpler. Typical CB units convert the 27-MHz signal first to an intermediate frequency (IF) in the 9–11 MHz range and then convert the resulting signal to a second kHz. Since this dual conversion process has advantages for selectivity and image rejection, it is highly desirable.

Many CB units employ an RF amplifier stage. This is a good feature for improving the unit's ability to receive weak signals. It increases the receiver's sensitivity.

The Cortez 40 uses a ceramic filter in the IF amplifier. The ceramic filter is useful for improving receiver selectivity. *Selectivity* is the ability to receive desired signals while rejecting others.

The unit transmits and receives on all 40 CB channels. The Cortez 40 employs *synthesis* to obtain the required frequencies and phase-locked loop circuitry to insure "on frequency" operation. Synthesis is a process of adding and subtracting frequencies to obtain other frequencies. With it, the transmit and receive frequencies for 40 channels can be generated from only a few crystal-controlled oscillators. *Phase-locked loop circuitry* eliminates most of the crystal-controlled oscillators by using one low-frequency, very stable oscillator to lock or stabilize noncrystal-controlled oscillators. The result is that this 40 channel unit has only one crystal and yet is more stable than traditional crystal-controlled units. Some units that provide for a lesser number of channels may not provide all of the crystals required. If crystals must be purchased, the additional cost may raise the price of the limited-channel unit to the level of a full-channel set which includes all crystals.

Sensitivity is an important measure of receiver performance. It is defined as the strength of a received signal which is required to provide a specified amount of signal level as compared to background noise. A radio which is receiving no signal will amplify the background noise in the form of hiss or static. When a CB signal is received, it is amplified. As the strength of the CB signal increases, it becomes louder in relationship to the background noise since the background noise remains constant in strength. The loudness difference between the signal and the background is expressed as a ratio measured in *decibels* (dB). One decibel is defined as the smallest change in sound level the human ear can detect. Table 2-5 gives a relative scale of different decibel readings. The VU or "S" meter of a CB radio is often calibrated in *volume units* (VU). A VU is equivalent to a dB except that the VU is more accurate for measuring voice and music while the dB is used for test tones.

A CB radio is very sensitive. Because the radio signals travel from the transmit antenna in all directions, the signal strength becomes less as it travels further from the antenna. By the time it reaches the receive antenna, the signal is measured in millionths of a volt or less. A millionth

TABLE 2-5   Signal-to-Noise Ratio

(S+N)/N	Description
1 dB (S+N)/N	The signal-plus-noise is only 1 dB greater than the noise. The signal and the noise are almost equal in sound level.
5 dB (S+N)/N	The signal is distinguishable above the noise, but the noise is very distracting.
10 dB (S+N)/N	The signal is very readable, but the noise can still be heard.
20 dB (S+N)/N	The signal is very clear; the noise, very weak.
30 dB (S+N)/N	Practically no noise is heard behind the signal (the signal is 1,000 times stronger than the noise).

of a volt is a microvolt (μV). The sensitivity of a radio is indicated by the increase in output it provides when a specified amount of signal is received versus the output it provides when noise only is received. This measure is referred to as the *signal-plus-noise to noise measure*: (S+N)/N. For AM reception, the receiver should deliver 10 decibels (dB) more of signal-plus-noise than noise alone for a received signal strength of 1 microvolt (μV) or less. The Cortez 40 provides 10-dB signal-plus-noise to noise for less than 1 μV received signal level. This indicates a very good sensitivity rating. The less received signal level required to obtain the same (S+N)/N figure, the better the sensitivity.

Intermediate frequency (IF) is not so important in itself. The important factor for AM units is that dual conversion is indicated by the presence of two IF frequencies. The frequencies will include one in the megahertz range and one in the kilohertz range.

Audio output determines the amount of audio power that will be available to drive the speaker. Two to three watts are normal and adequate. With an efficient speaker, this output range will usually deliver more volume than is required. It should be understood that the amount of speaker volume is dependent on the received signal level. If the station being received is too weak, increased audio output will only make the background noise louder.

Receiving current drain indicates how much of a load the radio will place on the power source when in the stand-by mode (not transmitting and receiving no signal). This is a significant measure for mobile, portable, and emergency rigs. The less the drain, the longer the battery will last. Thus a low drain figure is desired.

The Cortez 40 specifications do not identify the accessory functions provided by the unit. Comparison of supplied features is an important decision tool, however. Consideration of the sales literature determines whether a unit provides the following extras for use in the receiver mode:

1. RF gain control
2. Squelch
3. Delta tuning
4. Automatic noise limiter
5. Signal strength/RF power meter

The RF gain control is helpful when strong signals are present. The RF gain can be reduced so that background noise and weaker signals are not heard while the strong signal is left clear. If a weak signal is desired to be heard, the RF gain can be increased.

Adjustment procedures of the receiver squelch control were described earlier. Squelch makes normal channel monitoring more pleasant since it permits the background noise to be blanked out while stronger signals break squelch and are heard. While the receiver is squelched, no sound is heard from the speaker.

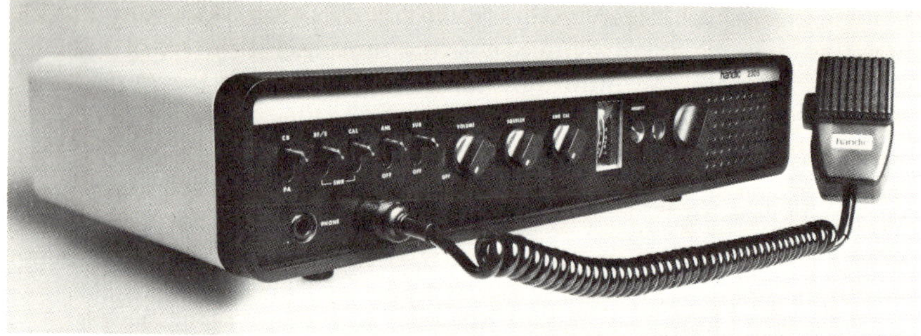

Fig. 2-9 A base station transceiver (*Courtesy*, Handic U.S.A., Inc.)

Delta tuning controls help to tune signals for maximum clarity. If the frequencies of different stations vary, the signal can be tuned for optimum quality. These tuning circuits affect the receiver tuning only, leaving the transmitter at its crystal frequency.

The Cortez 40 has a switch marked "NL" at one side and "Off" at the other. The noise limiter (NL) clips off the noise peaks so that the signal can be as loud as possible while the distraction caused by the noise is reduced. Some units provide noise blanker circuitry in addition to the NL circuitry. The noise blanker (NB) is more sophisticated and actually removes impulse noise. For mobile applications, the noise blanker is of particular benefit. Some units will permit either function to be turned on or off independently.

The Cortez 40 meter serves a dual function. It indicates the strength of the received signal, and it also indicates the relative level of RF power output during transmit. The received signal level is used to analyze received signals. The RF power output measure indicates how well the CBer's transmitter and antenna system are working.

The frequency specifications for the transmitter will normally be the same as for the receiver. Some units may provide special receive circuits such as the Handic base station, which has a second receiver that permits monitoring of channels different from the transmit channel (see Fig. 2-9). Unless special circuits or provisions are included, the frequency range of the transmitter should be the same as that of the receiver.

RF input and *output power* are important transmit specifications. The FCC limits both of these power measurements. AM input power is limited to 5 watts maximum, and output power is limited to 4 watts maximum. (Output power is the more important measure.) These two measures together represent the efficiency of the unit. The CBer should be sure that the unit will deliver 4 watts output with 5 watts input. If it will not, the output signal will be decreased in strength.

Adjacent channel rejection is a measure of receiver selectivity. It indicates how well the unit will be able to select one channel without receiving annoying interference from neighboring channels. The −60 dB figure indicated for the Cortez 40 is a very good rejection figure.

Modulation is limited by the FCC to a maximum of 100 percent. (See Chapter 5 for more details.) Also required is circuitry which will provide absolute limiting of modulation to no more than 100 percent. Ninety percent modulation is a typical reading indicating that a legal percentage of modulation is being maintained.

Transmitter current drain is evaluated similarly to the receiver current drain discussed above. The primary difference is that higher current drain is typical.

CB units will be powered by either 12 volts dc for mobiles or 120 volts ac for base stations. (Note: The dc voltage may be written 13.8 volts.) Some units provide for both types of power, which is a useful feature.

The specifications for the Hy-Gain V/Model 2705, which is shown in Fig. 2-10, are similar to those for the Cortez 40 except that SSB features are included and also a few other additions.

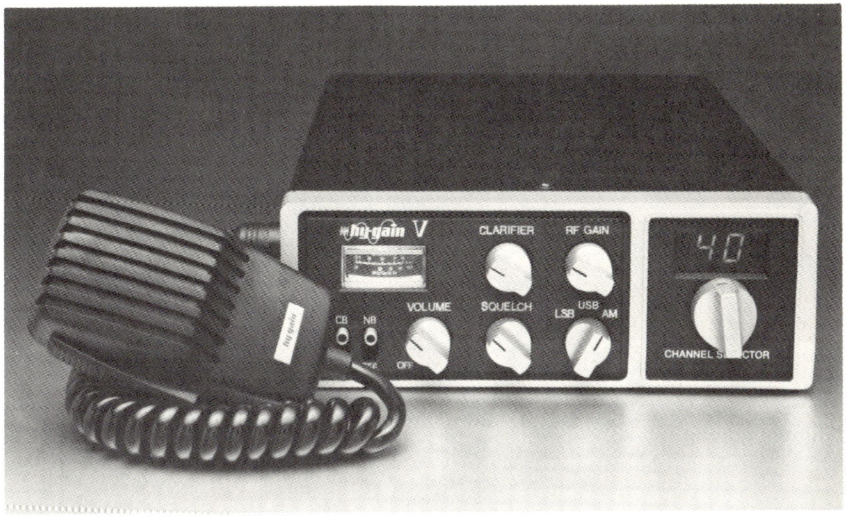

Fig. 2-10 The Hy-Gain V/Model 2705 CB transceiver. (*Courtesy*, Hy-Gain Electronics Corp.)

The "complement" specification identifies the type and number of active devices used in the unit. This specification may give some insight into the complexity of the unit, but it cannot be used for straightforward comparison of different units since one integrated circuit may be the equivalent of many transistors.

Squelch range is given in this specification, and this can be used for comparison with other units. Generally, the greater the range, the better.

Sensitivity includes an additional specification for SSB. It should be noted that SSB sensitivity is greater than AM sensitivity. Selectivity is also more stringent for SSB. Notice that the SSB selectivity figure has less than half the bandwidth of the AM figure (2.4 kHz vs 6 kHz). Power is given for both AM and SSB. The two power figures given represent the maximum allowable for each mode (see Chap. 5).

The receiver tuning discussed above for AM units is also available on the Hy-Gain V and is referred to as "Clarifier."

The SSB unit has several added specifications. In SSB units, the unwanted side band is rejected so that the desired side band is the only one transmitted. A rejection of 40 dB or more relative to the desired side band is desirable. The carrier is also suppressed in the SSB unit (see Chap. 5). For both unwanted side band and carrier suppression, the greater the suppression, the better.

Harmonic suppression must be at least 60 dB for units manufactured after September 10, 1976. This means that all undesired signals are reduced in level by at least 60 dB from the level of the proper signal. A signal which has a level reduced 50 dB from another is 1/100,000 as strong. A signal reduced 60 dB is 1/1,000,000 as strong.

The actual number of channels available to an SSB unit must be understood. Single side band (SSB) units use the same channels as AM units. In the SSB mode, however, the unit can operate on either the upper or lower side band of the channel. The result is that twice as many channel possibilities exist for SSB as are available for AM units. Some manufacturers add the number of AM channels to the number of SSB channels and advertise 69 channels for 23 channel units or 120 channels for 40 channel sets. Since SSB and AM cannot use a channel simultaneously without interference, a more accurate channel capability description would indicate how many AM channels are available and then indicate that upper and lower side band operation are possible in the SSB mode.

The Hy-Gain V has a public address feature. If an external speaker is provided, public address is made possible by flipping a switch.

The audio frequency response is given for the unit. Since only voice is used on CB radio, the response range can be less than that required for high fidelity record players.

When comparing CB unit specifications, the best procedure is to place the specifications side-by-side and evaluate one criterion at a time. If some kind of mark is devised to indicate which unit has the superior specification at each point, a quick check following comparison will indicate which unit has the larger number of superior features. From this information and all the other data that has been obtained, the CBer should be able to exercise his right of personal preference with assurance of making the best choice possible.

CHAPTER 3

Operating on the Citizens Band Mobile and Portable Service

Introduction to Mobile/Portable Communications

The previous chapter explained that a base station has the advantage of a greater range of communications than other units. It is limited, however, in its lack of portability. Mobile and portable Citizens Band equipment provides operators with the convenience of mobility. In a highly mobile society, this characteristic is greatly prized and is undoubtedly a major factor in the burgeoning interest in CB radio. Not only can a person remain in touch with others for business and domestic purposes, he can also enter into an interesting and fun-filled hobby. Time which before had been wasted in dodging fellow drivers and racing amber lights now is put to meaningful and enjoyable use. Without question, mobile operation is the most popular type of CB activity.

Mobile CB radios take many forms. Units are available which provide Channel 9 operation only. These sets are useful for emergency communications by motorists, but FCC regulations dictate that Channel 9 be reserved for these communications alone. The use of Channel 9 units is therefore very limited. Other mobile units provide three or five channel capability, and still others make provision for all 40 channels. Some mobile transceivers possess a level indicator meter. Others do not. Some provide for public address capability. Others have no such provision.

Just as is true with base stations, the kind of mobile equipment you choose depends primarily on the nature of your intended use of the unit and the size of your budget. Mobile units can be purchased for less than one hundred dollars. Single side band rigs may cost over three hundred dollars. Add multiple antennas, power microphones, VSWR meters, and similar accessories, and costs will rise accordingly.

Mobile radio has its own unique set of problems. The mobile CBer must divide his attention between operating the CB radio and driving the vehicle. Placement of the unit becomes an important consideration. Because one hand must be used for steering, simple operation is a definite

Fig. 3-1 *Control It from the MIC:* Mobile operation is made safer by CB transceivers which place the controls on the microphone (*Courtesy*, Hy-Gain Electronics Corp.)

requirement. The CB radio shown in Fig. 3-1 has all controls located on the microphone. The operator can adjust the unit to the proper channel, squelch, and audio level with one hand. The microphone control can be held at a level which permits viewing of traffic while adjustments are made. This precaution reduces the possibility of accidents.

Portable CB equipment is mobile equipment carried one step further. The portable unit is a complete, compact station. A built-in antenna is provided, and self-contained batteries supply power. Often the portable unit is designed for one-hand operation. The microphone and speaker are positioned so that the user can orient the antenna vertically while talking or listening. Such a unit is called a "handi-talkie" or a "walkie-talkie." Portable units are available in many varieties. Part 15 of the FCC Rules and Regulations defines and specifies an unlicensed class of CB radios, low-power units often sold as children's toys. "Kiddie-talkies," as they are called, are not well liked by licensed CBers. The reason is that they add confusion and interference to an already overcrowded band condition. FCC rule changes will alleviate many of these problems since this class of unit will be restricted to frequencies removed from the 11-meter band.

Licensed CB portable units (such as the one shown in Fig. 3-2) differ from unlicensed radios primarily in the power input, which ranges from 100 milliwatts to 5 watts, and the number of channels available.

The higher power units will, of course, deliver more consistent communications at a greater range. The range difference may not be very great, however. On flat, open ground, a range of one mile for 100-milliwatt units may increase to five miles for 5-watt units. This increase in range is gained at the expense of greater size, more weight, and shorter battery life. A difference of a pound or two may seem inconsequential, but, after several hours of holding the heavier unit to one's ear, the extra weight will seem large indeed.

Some portable units have provision for an external antenna and external power. Some units will even accept an external microphone with built-in thumb switch. These features mean that a unit used in the field as a portable set one day can become a base or mobile station another day. Some of these units may even surpass the standard base or mobile unit in operating efficiency.

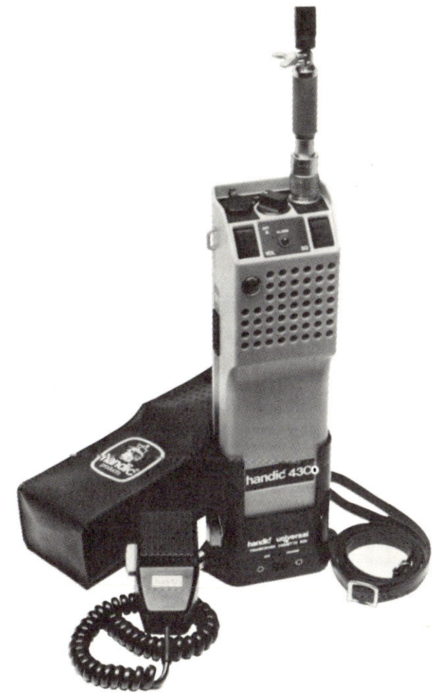

Fig. 3-2 A portable transceiver. (*Courtesy*, Handic U.S.A., Inc.)

Mobile Installation

How a mobile unit is installed depends on many factors. The type and size of the vehicle that it will be used in is important. The suggestions of the XYL should be considered. Ease of use and security must not be forgotten.

Probably the first decision to be made concerning the mobile installation is where to locate it. The most common location is under the dash. For the sake of appearance, the unit may be placed at the center of the dash facing straight back. This position may look the best, but, in any except the smallest car, it is not the most advantageous location for operating convenience. The CBer driving the car may have to take his eyes off the road, lean away from the steering wheel, and, even so, be unable to make fine adjustments on the unit without strain. It may be wise to sacrifice symmetry for safety and convenience. The unit can be moved closer to the driver and mounted with the front of the unit angled toward the driver for easier operation. If passengers wish to use the unit, they are free to move closer to the radio without jeopardizing the safety of the other occupants of the car.

If the radio is used primarily or exclusively by the driver, the unit can be mounted on the left side of the steering column where it will not only be close to him but more in line with his field of vision. This left-hand location is more advantageous for right-handers. The microphone is held in the left hand, and steering is accomplished with the right. In a vehicle with standard shift, the microphone can be held next to the steering wheel in the left hand while the right hand is left free to shift gears.

All under-dash mounting locations require similar installation practices. The radio is usually held in place by a mounting bracket bolted to the dash. Holes must be drilled into the dash, and if a mounting template is not provided with the unit, it would be helpful to make one. Light cardboard or heavy construction paper may be used for the template. The mounting bracket is laid down on the paper and traced. The mounting holes are drawn as accurately as possible. Once this is done, center marks are drawn in each hole, and the template is cut out. It is now the same size as the bracket but more manageable.

Before installation, you should procure safety goggles. The installation can become painful and costly if you have to go to a doctor as a result of metal fragments getting in your eye while drilling the dash. Figure 3-3 shows a typical set of useful tools for installing a CB radio in a vehicle.

Two practical steps will improve the chances for a successful installation. First, move the seat back as far as it will go. Second, carefully inspect the proposed location of the radio to make sure that nothing will interfere with the drilling. The area should be free of wires, pipes, switches, and similar hindrances. The projection of the radio should also be checked to make sure that the sharp edges of the unit will present as

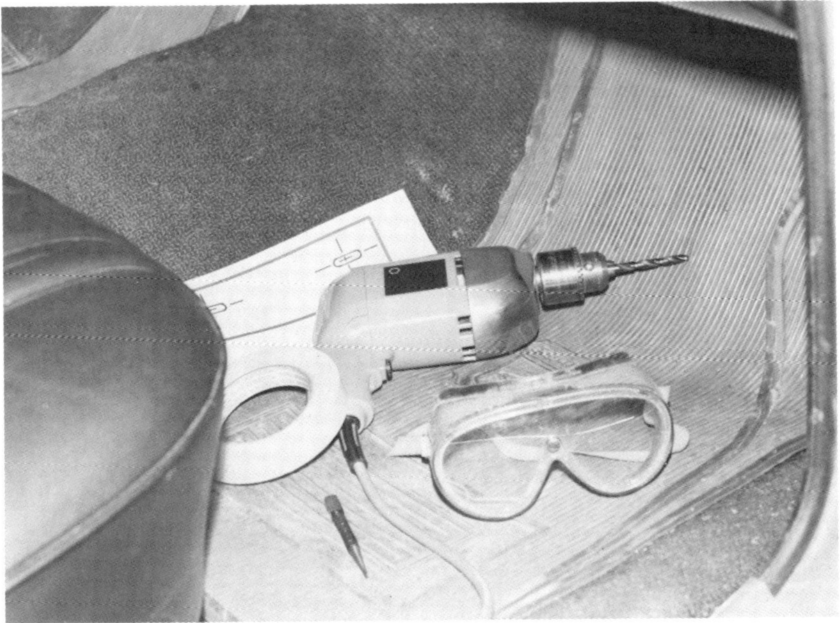

Fig. 3-3 The tools of mobile installation.

little danger to knees as possible. If the car uses a floor-mounted shift lever, be sure that the radio will not interfere with its operation.

With these precautions taken, the installation can begin. First, the template is positioned in the exact location decided upon as best. It is then taped to the dash in such a way that its position does not change. A center punch is placed at the cross made in each hole. The punch is tapped hard with a hammer to create a small depression in the dash. When all of the holes to be drilled are punched, the template is removed.

The installer should then put on the safety goggles and situate himself as comfortably as he can. The drill bit should rest in the depression made by the center punch. If a variable speed drill is used, the drilling should be started at a low speed and with a light pressure so that if the bit does slip out of the hole, the finish of the dash will be less likely to be damaged.

Nuts and bolts or self-tapping metal screws can be used to mount the bracket. The hole which is drilled, of course, must be of an appropriate size for the self-tapping screw or bolt. Lock washers should be installed if required.

To increase stability and reduce vibration, a metal strut can be installed at the back of the unit. This strut should attach to a screw on the radio and an existing screw somewhere beneath the dash, or a self-tapping screw can be placed in the fire wall to hold it. It is wise to remember to "look before you drill" when putting this screw in the fire wall.

Second in popularity to the under-dash mounting location is the hump mounting location. Mounts similar to the one shown in Fig. 3-4 can be obtained which will permit CB radios to be installed on the transmission hump in the center of the floorboard between the dash and the front seat. These mounts elevate the front of the unit so that its faceplate is readily observable by the driver of the car. The unit is handy when installed in this location. It is nearer to the driver than when it is mounted under the dash, and it can be seen easily. The driver's eyes are diverted from the road when he looks at it, but only momentarily, since the unit is in clear view.

Fig. 3-4 A transmission hump mount for a CB radio with speaker. (*Courtesy*, Acoustic Fiber Sound Systems, Inc.)

Requirements for installing a CB unit on the transmission hump depend on the mount selected. Specific instructions will be provided by the manufacturer. Generally speaking, hump mounts require attachment to the floorboard of the vehicle, usually by means of self-tapping screws. One problem which may be encountered is the tendency of the threads of the carpet and carpet pad to wind on the drill bit and impede the drilling. To prevent this, one can use a sharp pen knife to cut a small slit where the screw will be installed. A short piece of copper tubing, which should have an inside diameter greater than the diameter of the drill bit, can now be placed into the slit. The tubing is held with pliers, and the drill bit is placed inside it. After the hole has been drilled, the tubing is removed. Alternatively, it may be possible to hammer an ice pick or nail through the floorboard and to use its penetration as the pilot hole for the self-tapping screw. If this proves possible, drilling will not be required. One should check under the floor to be sure that nothing is in the way before making holes by either method.

Mounting the radio on top of the dash has several advantages. Primary among these is that the location places the unit closer to the driver's normal field of vision. His eyes must move from the road only slightly to see it. Indeed, any sudden movement on the roadway will probably enter the driver's peripheral vision.

Two major drawbacks plague the dash-top mounting. It is difficult to route cables to a unit thus mounted and especially to do so neatly. A second problem is that the unit is in plain view of all who walk past the car when it is parked. Considering the large number of CB radio thefts, advertising the presence of a CB unit in this way may not be wise.

Installation procedures for dash-top mountings are basically upside down from those for under-dash mountings. The same mounting bracket will hold the unit on top of the dash by merely being turned over. Self-tapping screws secure it to the dash top. It is important to place the bracket so that the screws will have metal to screw into. If the dash is padded, the screws might miss the metal altogether and be installed in the padding only. Special caution should be exercised to insure that nothing is mounted under the dash at the point that the holes are drilled.

A variation of the dash-top mounting is the overhead installation shown in Fig. 3-5. The same convenience versus security trade-off applies to this location. The ceiling mount, however, has gained popularity among owners of vans.

Some CB radios are small enough to permit their installation in glove compartments. The benefit in terms of security is self-evident. The inconvenience for operation by the driver of the car is also apparent. The CBer must weigh benefits and liabilities to make the final decision on glove compartment installation. Since such installations are usually temporary, no permanent attachment is required. Once the cables are hooked up, the unit is simply stowed in the glove compartment. Only the microphone is taken out during use.

Some recent CB units are divided into several parts located in different areas of the car. The control head of remote control radios is mounted in the same manner as a typical CB radio. This control unit is not a complete transceiver. It connects with a second unit which can be located in the trunk or some other secure area. The major benefits are

Fig. 3-5 Ceiling-mounted CB unit. (*Courtesy*, J.I.L. Corporation of America)

twofold. The unit is more protected from theft, and the energy lost in the antenna cable is reduced if trunk or rear-mounted antennas are used.

Power Connection

Power for the mobile CB radio is obtained from the automobile battery. There are several ways to obtain this power (see Fig. 3-6). Many CB units are supplied with a power lead containing an in-line fuse. This lead can be connected directly to the battery if desired. The fuse will protect the unit, and the power will be available at all times. Direct connection to the battery will permit the unit to be operated without the motor running and even without the key inserted. This situation is undesirable for several reasons. If the radio is left on and the car is unused for a long period of time, the radio could run the battery down. Also, with the radio hooked directly to the battery, there is no way to safeguard the use of the radio. If children or unauthorized individuals transmit over the radio in illegal ways, the licensee would be held responsible.

It is possible to connect the CB radio to a terminal on the rear of the ignition switch so that power will be supplied only when the key is switched on. This arrangement permits the radio to be operated only by persons having the car key. It also insures that when the car is not occupied, the radio will be off. The CB set is secure against unauthorized

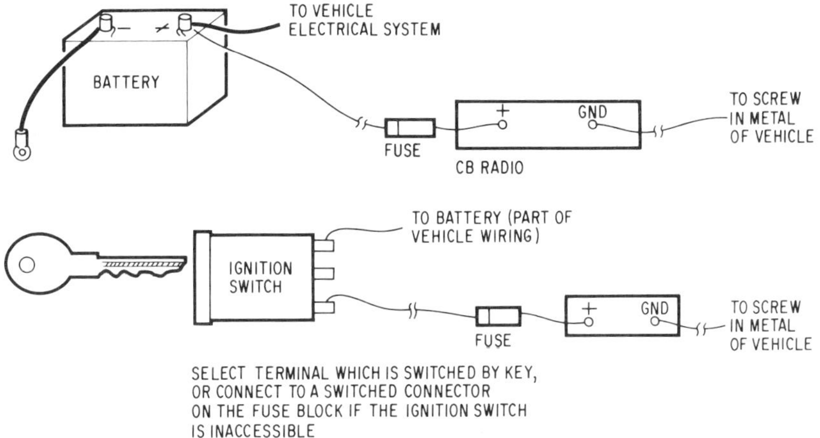

Fig. 3-6 Power wiring for mobile installations.

use, and the battery is protected from extended power drain. In some cars, the ignition switch is built into the steering column. To hook a wire to such a switch would require a master mechanic. Switched voltage is usually available at the fuse block, however. If the key switch is inaccessible, a wire can be routed to the fuse block and connected to a switched terminal. The car radio is often connected to a switched terminal, and wiring for the CB radio can be paralleled with it. Solder or crimp type terminals should be used for all voltage connections.

The CB radio should be well grounded to the car. Although the negative side of the battery usually connects to the radio via this ground connection, some cars may have opposite polarity, with the positive side of the battery connected to the frame ground. It is important, therefore, that the ground connection be properly made. Often, the bracket will ground against the dash. With the increasing use of plastics, however, bracket mounting may not provide a good connection. A wire can be connected between a screw on the radio and a convenient screw at the fire wall or some similar metal surface. Such a grounding procedure is highly desirable.

Most CB radios operate from 12 volts. If your car or boat has a 6-volt system, use of the CB radio may require a power inverter to boost up the 6 volts from the battery to 12 volts. The connection is simple. The input wires of the inverter are attached to the battery with the "+" or positive lead (usually red in color) on the positive terminal and the "−" or negative lead (usually black in color) connected to the negative terminal. The inverter output leads are attached to the radio, again observing the proper polarity. Once connected, the radio will receive its required 12 volts, and normal operation will be possible.

Mobile Speakers

Mobile CB units usually have built-in speakers. For many applications, these "in-board" speakers are adequate. Unfortunately, internal speakers are usually small because of the space limitations of the radio. The use of a small speaker results in the loss of bass frequency response. Other problems with built-in speakers are the lack of a proper baffle and the direction of the sound resulting from the location of the speaker.

For a speaker to operate best, the air movement caused by the front of the speaker must be isolated from the air movement caused by the back of the speaker, or it must be phased so that the sound waves caused by the two sides of the speaker reinforce each other rather than cancel each other out. A baffle performs this function.

Some baffles are constructed so that the air movement caused by the rear of the speaker is absorbed within the enclosure. The result is that no sound cancellation can occur. Other baffles are constructed so that sound waves from the rear of the speaker are reflected and ported back to the front of the baffle at an interval designed to cause addition of the sound waves, thus reinforcing them. High fidelity speaker enclosures are often built this way in order to enhance the bass frequencies. Understandably, a speaker crammed in among the wires and components of a CB radio can expect little help from its enclosure.

A second problem with built-in speakers for mobile use is that they are usually mounted either at the top or bottom of the unit. The sound energy is radiated in the direction to which the speaker points, that is, downward into the carpet of the car or upward under the dash. The operator must rely on reflected sound. To compensate for this low audio, he increases the volume, thereby causing distortion.

An external speaker is the logical solution to these problems. The external speaker can be larger; it can be enclosed in a baffle; and it can be mounted in a position which will provide more efficient operation. A speaker for CB use does not need to be gigantic. Large speakers increase the quality of bass frequencies only. Since human voices do not range below a few hundred Hertz (cycles-per-second), a speaker with 10 to 15 square inches of surface area will reproduce voice frequencies adequately. The baffle is more important than the speaker size. Baffles for CB speakers vary from simple boxes to specially designed acoustical absorbers.

A simple speaker box can be effective if it is properly located. Placement should be chosen by actual tests. The speaker should be temporarily located at the desired mounting position. The sound of the speaker should be evaluated for clarity and ease of hearing. The speaker should now be moved toward and away from the surface behind it and angled in several directions. Evaluations of each position should be made. One of these positions should provide the best sound with the least distortion and reflection.

Generally speaking, the speaker will give the best results when it is at the same height as the operator's ears, when it points directly at the operator, and when it is close to the operator. Such a position will permit the maximum sound energy to reach the ear when the speaker is operated at a lower level. Moreover, less reflected sound will be present. A speaker mounted on the top of the dash or on the door post of a sedan will work very well.

If an absorbing baffle is used, the placement is less critical. The speaker should still point at the operator, but its position relative to reflective surfaces will be less important. A well designed baffle will operate best at voice frequencies and will act as a type of noise filter, reducing "out-of-band" sounds while amplifying the desired sounds.

External speaker terminals or jacks are provided on most CB radios. It is desirable that the internal speaker be disconnected when an external speaker is used. This can be accomplished automatically by a special jack in some units. When the external speaker is plugged in, the internal speaker is cut off.

Wiring of the speaker consists of nothing more than connecting the external speaker terminals on the radio to the terminals on the speaker or baffle. Screw terminals may be provided, or solder lugs may be employed. If two speakers are used, phasing becomes important. Both speakers must push air at the same time. If one pushes and the other pulls, cancellation will result. Speakers are usually marked with a dot or a sign indicating how to hook wires to them. The wires connecting to the "+" terminal on the radio should be connected to the "+" terminal of the speakers. The ground or "−" terminal of the radio should connect to the other terminal of the speakers.

Cable Installation

The radio is mounted; the speaker is in place; the location for the antenna has been chosen. The problem now is to connect the various pieces of equipment together with wire. It is important to assure maximum performance while maintaining proper appearance. While the location of the radio and antenna will determine the specific methods used for mounting, some general methods apply to all installations. Mounting the radio in most positions will require only a small hole and discreet routing of the cable. Care should be taken to insure that cables do not block vents or foul fan blades. Wires run in the motor compartment should be kept clear of hot or moving engine parts. Cable clamps or tie wraps should be used under the dash to hold the cable away from the legs of the driver and passengers.

At the radio end of the cable a connector must be installed. A PL-259 connector is the most common type used. Different types of connectors are designed for different types of coax. The CBer should make sure he

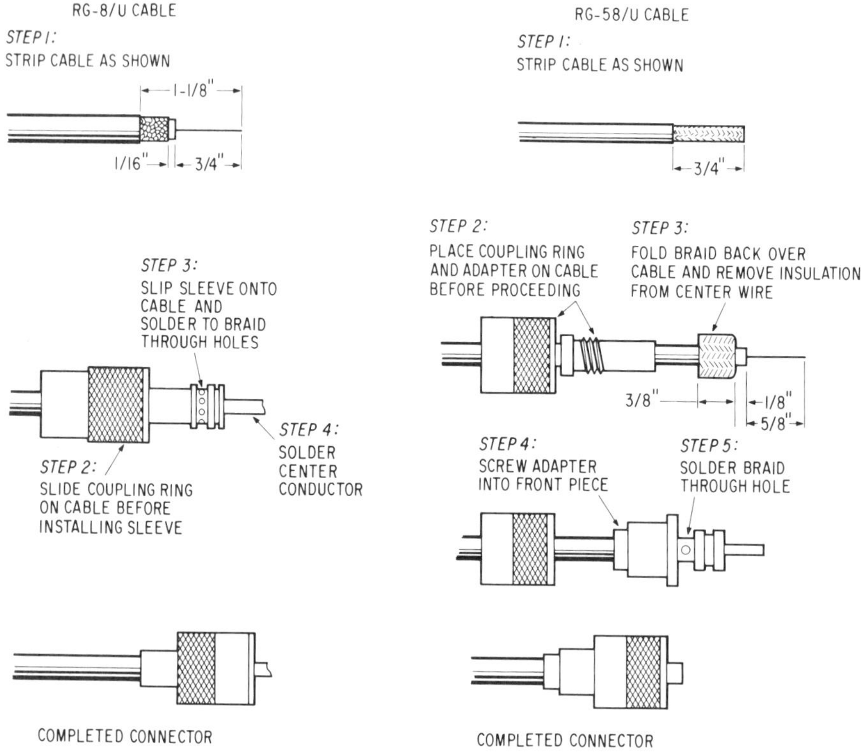

Fig. 3-7 Coaxial cable connector installation.

matches connector and cable. Figure 3-7 shows how to install a PL-259 connector.

The antenna cable will usually extend from the front of the car to the antenna mounting position. Some differences in cable routing will be necessitated by different antenna mounting locations. For roof-mounted antennas, the cable is first routed beneath the carpet. Depending on the type of car that is involved, the cable may be routed to the roof area at the front or middle of the vehicle. A sedan has a roof support between the front and rear door. This support is hollow and usually has cutouts on the inside surface, as shown in Fig. 3-8A. These cutouts are covered by decorative cover panels. It will be necessary to remove the door plate to place the cable beneath the carpet (see Fig. 3-8B). After routing the cable past the front door, it can be run into the roof support and up this support to the roof of the car. At this point, the cable must be run over the headliner to the point where the antenna is to be installed.

Since a coupe does not have the center roof support, the cable must be routed to the roof at a different point. If the front roof support is accessible, the cable can pass through it to the roof. If it is not, there may be room behind the decorative cover panels for routing the cable. If

Fig. 3-8 *Cable Installation:* (A) Hollow roof support with decorative cover removed, through which cable is routed to roof-mounted antenna, and (B) routing cable beneath door threshold.

neither of these methods is feasible, cable clamps can be screwed onto this front roof support and used to hold the cable in place. The cable will be exposed when this method is used, but if care is taken, the installation can be made neat and acceptable.

If a trunk or bumper antenna is used, the cable will need to be routed to the rear of the car. It should be placed under the carpet at the side of the car as described above. Instead of stopping at the roof support, however, the cable must be run into the trunk of the car. This is most easily accomplished by removing the bottom of the rear seat. The cable can be brought out from under the carpet into the area usually covered by the rear seat. A hole must be made into the luggage compartment from this point. The design of the car will determine the proper place for this hole. It can be drilled from the rear seat area or from the luggage compartment and should be just large enough to facilitate insertion of the cable. In some cars, the back of the seat is exposed in the luggage compartment. If this is the situation, it may be possible to push the cable up behind the back of the seat and into the luggage compartment without making any holes.

A hole is required at the rear of the luggage compartment to route the cable to a bumper-mounted antenna. The location of this hole should be chosen to prevent the entry of water into the luggage compartment. If the hole is made in the bottom of the compartment, some plumber's putty can be used to seal it off after the cable is installed.

Mobile Antenna Installation

There are several ways to mount antennas on vehicles. The grandfather of mobile antenna mounts is the adjustable ball and spring mount. This assembly is often used with long whips. The ball is constructed in two parts, with the division cut at an angle to the mounting shank and spring stud (see Fig. 3-9). By varying the way that the two halves fit together, the angle which results between the stud and the shank can be varied between zero and 90 degrees. This range permits a very flexible method of mounting. For mounting on the side of the vehicle, the ball is set for a 90-degree angle. The mounting shank is inserted into the bakelite base and is oriented to set the antenna straight up and down. If the mount is used on the top of the trunk deck, a zero-degree angle is used. For tilted surfaces, intermediate angles are used.

The mount is installed by drilling the car body to match the bakelite mounting plate. Often a template will be provided to assist in laying out the required holes. Once the location for the antenna is chosen, the template can be taped to the car body with masking tape. A punch should be used to make indentations at the crosses on the template which indicate the center of the holes. Note once again that the maxim for the mobile antenna installer is "look before you drill."

The installer should be sure that the chosen location is accessible

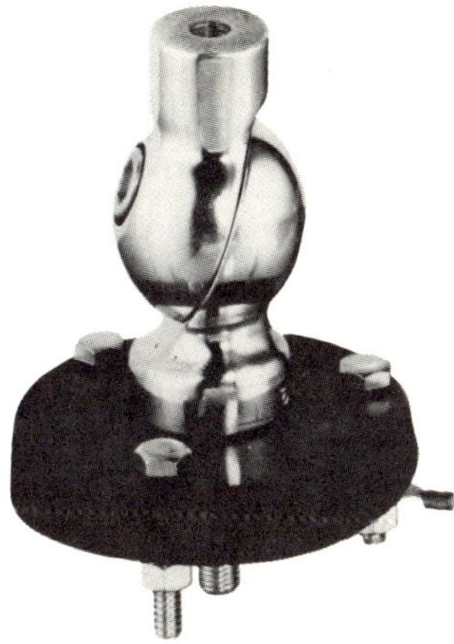

Fig. 3-9 Swivel ball mount. (*Courtesy*, Avanti Research and Development, Inc.)

from inside the trunk or fender and that pipes, wires and other items are not in the way of drilling. The preferred hole size will be indicated on the template or in the base installation instructions. The bolt holes can usually be drilled with a standard twist drill bit. The center hole will usually be larger than available drill bits. Hole saws can be used to make this larger hole. A metal punch, as shown in Fig. 3-10, can also be used to make large holes. A $\frac{3}{8}$-inch hole is first drilled at the center "x" of the template. The screw of the punch with the die on it is placed in the hole, and the cutter is screwed on the screw under the metal. When the screw is turned with a wrench, the cutter is pulled through the metal, leaving a large smooth hole.

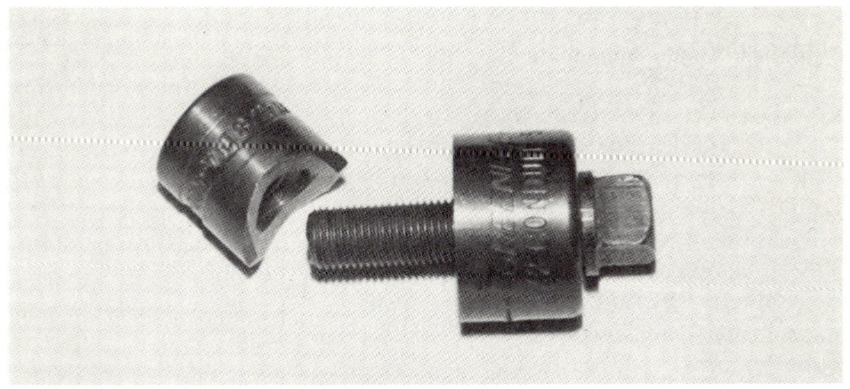

Fig. 3-10 Chassis punch.

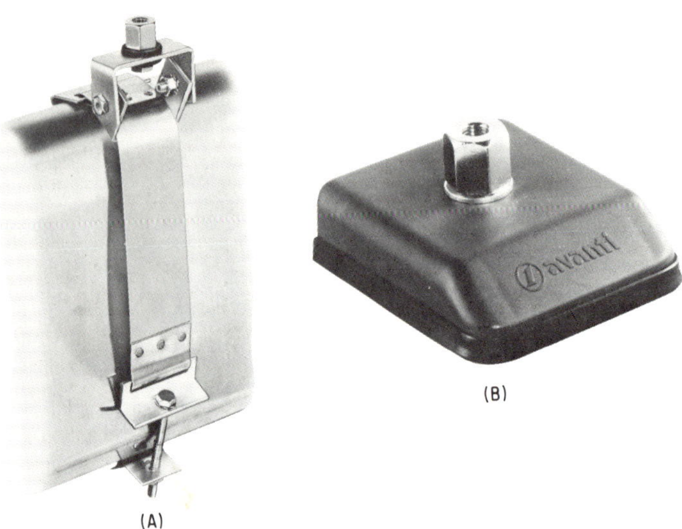

Fig. 3-11 *Mounts for CB mobile antennas:* (A) Bumper mount, and (B) trunk lid mount. (*Courtesy*, Avanti Research and Development, Inc.)

Mounts installed on a curved surface are subject to leakage unless protective measures are taken. Extra rubber spacers can be used to seal off the base against the car body. If no extra spacers are provided with the base, they can be cut from an old inner tube.

The coax is connected at the under side of the mount. A bolt screws into the center of the mounting shaft. The center conductor of the coax should be attached to an "eye" terminal lug, and this lug should be placed under the bolt head with a lock washer. The braid of the coax should be connected to a terminal lug, which is placed under the nut of one of the base mounting bolts. A cable clamp can be used to hold the coax at the base mount and thus prevent the center conductor from being subjected to stress caused by the weight of the coax.

The spring on this mount insures the safety of the whip. If low objects are struck with the whip, the spring will bend and leave the whip undamaged. It should be noted that the spring and base add to the length of the whip. Therefore, when a whip is chosen, it should be only 102 instead of 108 inches long.

If the antenna is to be bumper mounted, a special bumper mount can be obtained which is predrilled to accept the ball and spring base. Several types of bumper mounts are available. Most of them are universal mounts that are designed to fit bumpers of many kinds of cars. The bumper mount shown in Fig 3-11A uses a flexible metal strap which clamps to the upper and lower edges of the bumper and can be increased or decreased in length until it fits the bumper. One very flexible and sturdy mount uses two chains of link construction. This design permits the mount to be attached to cars with oddly shaped bumpers since the chains bend to fit the bumper's contours.

Coax connection for bumper mounts is the same as for body mounting. Only the routing of the coax differs. It may be necessary to run a ground connection.

Coil-loaded antennas are usually designed for convenience of mounting. One example is the trunk-mounted antenna shown in Fig. 3-11B. The mounting bracket for such antennas is designed to fit over the lip of the trunk lid. Set screws are provided to hold the mount in place. Installation is simple and attractive. Some antennas even have bases which match the color of the car they are used on.

As shown in Fig. 3-12, the location of a mobile antenna determines the pattern of its radiation. Short coil-loaded antennas may be installed at the center of the roof, usually with a feed-through mount. The coax

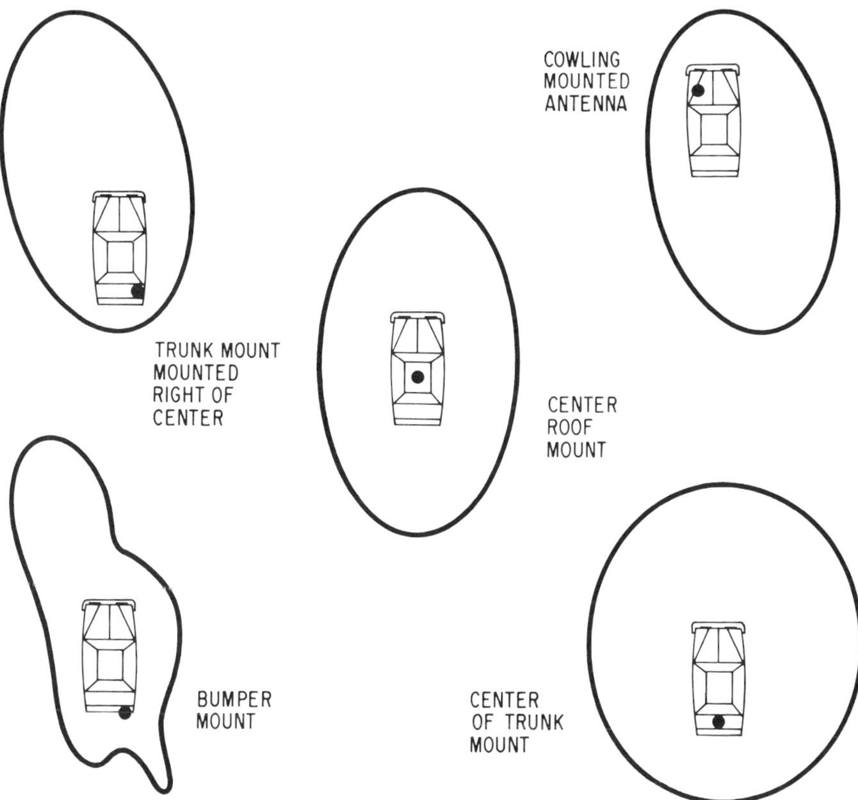

Fig. 3-12 Mobile radiation patterns vs. antenna placement.

connects to the bottom of the feed-through bushing, and the antenna connects to the top. Installation is greatly simplified if the antenna is placed directly above the dome light. The dome light is temporarily removed, a hole is punched and drilled, and the feed-through is installed. Getting the coax to the bottom of the feed-through can be somewhat tricky. As indicated earlier, the coax may be run up the hollow roof supports or hidden behind the decorative cover panels. Once the coax is at ceiling level, it must be run over the head liner to the dome light hole. It is then easy to attach the coax to the mount and replace the dome light. Such an installation is clean and efficient, but caution should be exercised to prevent tearing the head liner.

Some mobile antennas can be mounted without drilling holes in the vehicle. One of them uses a rain gutter mount that clips to the rain gutter of the vehicle (see Fig. 3-13A). Some semipermanent rain gutter mounts

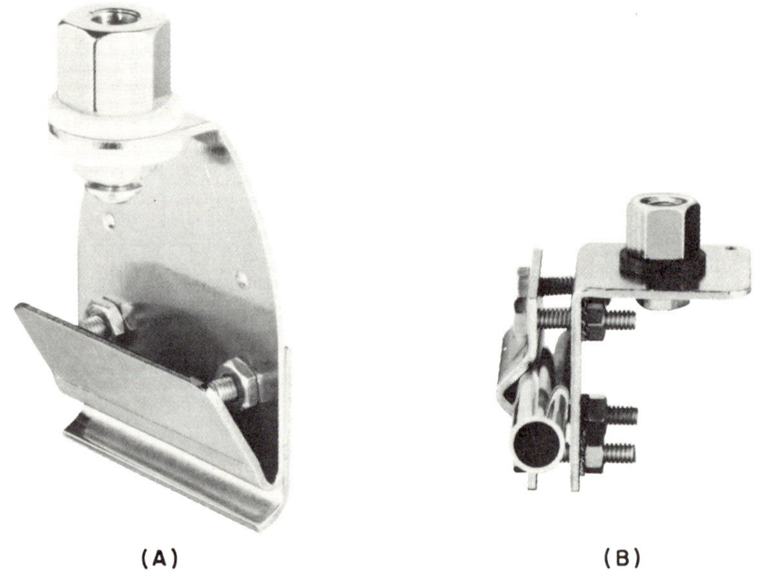

(A)   (B)

Fig. 3-13 *Mounts for CB mobile antennas:* (A) gutter clip mount, and (B) mirror clip mount. (*Courtesy*, Avanti Research and Development, Inc.)

are screwed on, but other mounts are merely attached temporarily with a spring loaded clip. The coax to these antennas is usually routed out a window or through the door. If the coax is crimped too hard by the window or door, the antenna's efficiency may be reduced and its visual standing wave ratio increased.

An antenna with a pipe mount can be used for vehicles with large mirrors supported by pipes. Because of the increasing number of CB-equipped recreational vehicles and the large number of truckers now using CB radio, this type of antenna has become particularly popular. Installation is easy and requires a minimum of tools. A saddle clamp is connected to the mount with screws, and the screws are tightened until the clamp holds the antenna firmly in place (see Fig. 3-13B).

The simplest antenna mount requires no installation at all. A magnetic-base antenna mount can be placed on any smooth ferrous surface and will remain there even at high speeds and over rough roads. For all that, it can be easily removed. In this day of widespread CB radio theft, the magnetic-base antenna has a lot of selling points.

Operating Portable Units

Portable units are self-contained, carrying their own microphone, speaker, antenna, power supply, and radio circuitry. Ordinarily, no installation is required, but the operator of portable equipment must work within its capabilities. It is usually not possible to add hi-gain antennas or other devices to the unit. Before purchasing one, the operator should be sure that portability is necessary for his purposes.

Portability is frequently an asset. When people wish to maintain contact over relatively short distances, a portable unit such as the one shown in Fig. 3-14 can provide the required mobility. A rancher who must

Fig. 3-14 Portable transceiver. (*Courtesy*, Hy-Gain Electronics Corp.)

contact ranch hands who are on horseback or a warehouseman who must be constantly alert for orders finds the portable unit a very useful tool. Sportsmen use them for weather reports and to avoid accidents. Construction workers often employ walkie-talkies to extend the range of their voices. Shouting and signaling can be replaced by conversational speech.

Portable units are not the answer for all problems, however, being limited in range and power supply life, and users must work within these limitations. If long distances or extended periods of time are involved, special solutions must be sought or some other system used.

Since portable units cannot be improved by adding accessories, the user must seek other ways to use them most effectively. Although it is not possible to increase their power output, it *is* possible to maximize their effective range. For example, in hilly terrain, the user of a walkie-talkie should maintain line-of-sight contact with other units if possible. This will mean that if one or more units must descend into a valley, at least one unit should maintain a position on a hill top which looks down into the valley. Otherwise, communication will probably have to cease until the units can return to the hill top. Search and rescue operations often work in this kind of situation. A mobile or portable command post is set up on the highest hill, and secondary units then act as relays between the command post and units in the valleys. Line-of-sight is the important rule for field operations.

CB radio can be used within and between buildings. Wood frame construction will attenuate the signals very little. Buildings of metal construction will attenuate signals in proportion to the amount of metal present and its proximity to the CB antenna. Glass does not hinder CB radio signals greatly. Signals travel farthest over an unobstructed path, but communication may be maintained over shorter distances despite obstructions. Communication can often be established or improved by moving a unit a short distance so as to avoid the interference of metal or other signal-blocking barriers.

For units transmitting over level, open ground or over open water, the communications range is limited by the curvature of the earth. Radio waves travel in a straight line, and although they may "bend" a little and thereby increase their range, there is a radio "horizon" beyond which they cannot follow the surface of the earth. That is why towers are used for antennas at base stations and why portable users must often climb hills or trees.

A second limitation of portable operation is battery life. The portable unit carries its own batteries, either standard flashlight batteries or the rechargeable nickel-cadmium type. The initial cost of the latter is greater, but in the long run they are more economical. Whichever battery is used, the operator should try to extend its life. Transmissions should be kept as short as possible and made only when necessary. Units use much less power in the receive mode than when transmitting. If a transmit range switch is available, it should be set to the lower power setting whenever possible.

Battery drain can be reduced during the receive mode also. Speaker volume should be kept as low as possible. An ear plug will help cut power consumption. Use of a quiet channel avoids the power waste of listening to other stations. Even with these procedures, a spare set of batteries is a wise precaution.

It was mentioned earlier that portable units can be obtained with features which permit their use as mobile or base units. Consideration should be given to such units because of their flexibility. Power output and channel capacity are also factors to be considered in choosing a portable unit.

CHAPTER 4

Antennas for the Citizens Band

Antenna Basics

The Class D Citizens Radio Service is assigned to the 11 meter band of frequencies. The designation, "11 meter," refers to the wavelength of a single cycle of energy. Wavelength is directly related to the frequency of the energy. A radio signal is electrical energy which varies in strength, changing from a minimum value to a maximum value and back to a minimum value.

The action is not unlike a rolling wheel (see Fig. 4-1). If a mark is made at the bottom of the wheel, and the wheel is then rolled until the mark is at the top, and then rolled further until the mark once again is at the bottom, the mark will have made a complete circle. It has proceeded from the lowest position to the highest position and then back to the lowest position. Notice that it has also traveled a certain distance in the direction in which the wheel was rolled. This distance is the distance the wheel travels each time it completes one cycle. If it is rolled fast enough to complete one cycle each second, it will travel this distance each second. The distance between the lowest and highest points of succeeding cycles can be referred to as its wavelength. The wheel can be said to have a frequency of one cycle per second and a wavelength equal to the distance it travels in one second.

A radio wave also travels in a circle. It starts from its lowest value, increases in strength until it reaches its highest value, and then decreases until it reaches the lowest value again. If the radio wave returned to the point where it started, it would circumscribe a circle. Like the wheel, however, the radio wave travels a set distance during the time it changes values through each cycle. The distance that the radio wave travels in the length of time required for one cycle is defined as its wavelength. For radio waves in the Citizens Radio Service, channel 11 for example, the wavelength is based on the speed of electromagnetic radiation, which is 300,000,000 meters per second in free space. Since channel 11 has a frequency of 27,085,000 cycles per second (the unit of cycles per second is now called Hertz), a wavelength is equal to

58

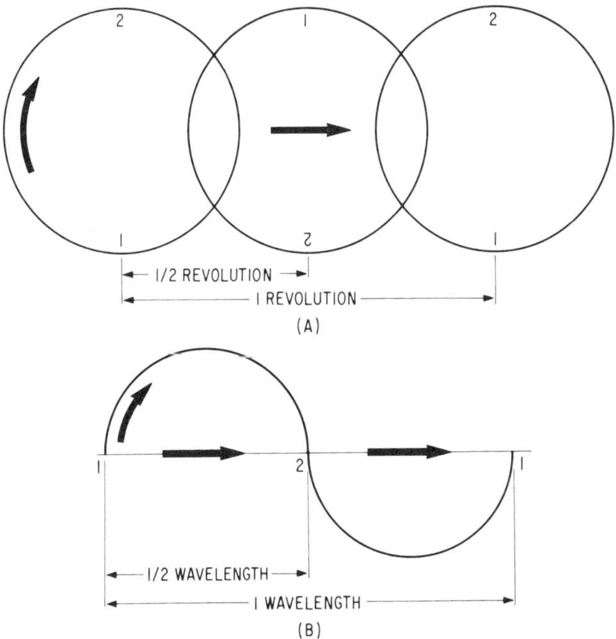

Fig. 4-1 *Wavelength of radio frequency energy:* (a) The wheel starts with the "1" at the bottom, turns until the "2" is at the bottom, and continues until the "1" is again at the bottom. The wheel has traveled a set distance equal to its circumference. (b) The electrical wave makes a complete cycle. It travels a distance equal to one wavelength.

$$\frac{300{,}000{,}000 \text{ meters/sec}}{27{,}085{,}000 \text{ cycles/sec}} \text{ , or 11.076 meters/cycle}$$

The length of a cycle of radio frequency energy is important in antenna design. For an antenna to be most effective, it must radiate when it is transmitting and absorb as much energy as possible of the frequency being used when it is receiving. To do so, the antenna must have a length which is related to the wavelength of the signal being used. An antenna which is used for channel 11 operation should thus have a length related to 11.076 meters (36 feet, 4 inches). It could be a multiple of this length or a fraction of this length. The important consideration is one of signal phase. One cycle contains all values of a signal from zero to maximum. If the signal reaches the receiver from two different paths so that a minimum value from one cycle and a maximum value from a second cycle are paired up, cancellation will result. If, on the other hand, the maximums of two different cycles reach the receiver at the same time, they will add, and the signal strength will increase. The antenna provides this extra path. To reach the receiver, the energy must travel down the antenna. The time required for the energy to travel down the antenna is related to the wavelength of the radio wave cycle. An antenna of the proper length will pair up a maximum from one cycle with a maximum from a second cycle. These two maximums are in phase and therefore combine. Other signals with different wavelengths will be cancelled because of the antenna length. In this way the antenna is able to select and amplify desired signals and reject or reduce the level of other signals.

Experience has demonstrated that an antenna does not need to be a full wavelength long to operate properly. Most CB antennas are fractions of a wavelength. Antennas one-half or one-quarter wavelength long are often used.

Antenna Types—Base Station

Antennas for use on the Citizens Band can be obtained in many shapes and sizes. Some are simple and inexpensive. Some are gargantuan in size, complex in construction, and prohibitive in cost. Regardless of the type and size, the purpose of all antennas is the same—to establish communication.

When antennas are compared, some reference must be used. The reference used by antenna manufacturers and engineers is the *isotropic radiator*. An isotropic radiator is a theoretical antenna. It is conceived of as an antenna which radiates energy from a point in space in all directions at the same time and with the same intensity (see Fig. 4-2). The waves radiating from this antenna form a perfect sphere that grows progressively larger the farther the waves travel from the source. Since the energy cannot be increased once it leaves the antenna, the effective strength of the signal is proportional to its distance from the antenna. For each unit of distance the energy moves from the antenna, it must cover an area increased by the square of that distance. Therefore, if the signal is one foot from the source, it covers an area of one square foot. At two feet from the source, the area is equal to four square feet. At ten feet, the area is one hundred square feet, and so on. The strength of the received signal at each distance from the source will diminish by the reciprocal of this proportion (that is, 1/4, 1/100, etc.).

The half-wavelength dipole is a simple antenna which is often used as a standard for antenna comparisons. The dipole in its simplest form is not often used by CB operators, the primary reason being that the dipole is a horizontal antenna and CB antennas are usually vertical. There are times and places when the dipole is useful, however. The one shown in Fig. 4-3 will operate very well on the Citizens Band. It is a good antenna for emergency operation because of its ease of installation. A dipole can be constructed and stored in a closet for emergency use, for example, when other base station antennas are destroyed in a storm.

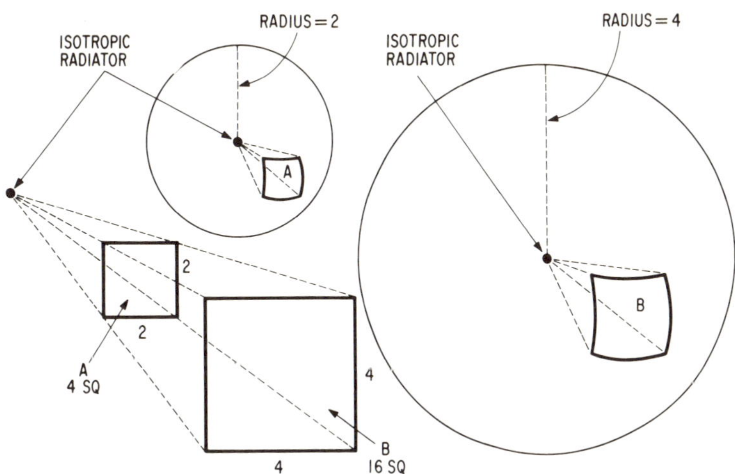

Fig. 4-2 *Energy spheres of an isotropic radiator:* The two spheres represent the radiated energy of radii 2 and 4, respectively. The quantity of the radiated energy is the same, but the second sphere is larger. The squares drawn on the surface of the spheres demonstrate that the area which the energy covers is equal to the square of the distance from the source. An antenna at the distance of the larger sphere will receive ¼ the energy that would be received at the distance of the smaller sphere.

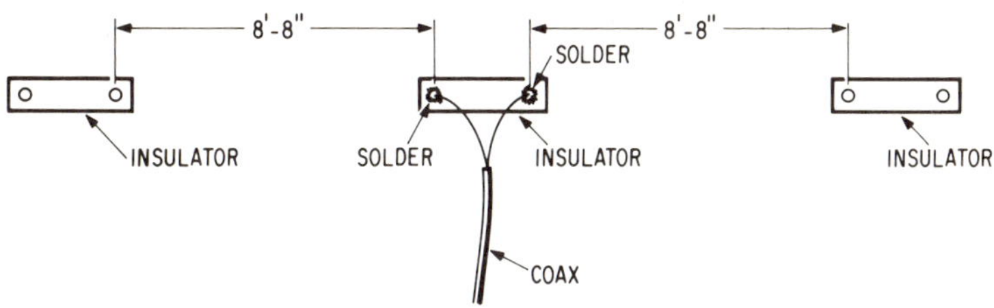

Fig. 4-3 *Dipole antenna:* A dipole constructed as shown is handy for emergency operation.

The dipole has a broad directional pattern. It presents approximately 2 dB of gain over the hypothetical isotropic radiator because it limits the pattern of its radiation. As indicated above, the dipole is horizontally polarized. This means that its electric lines of force run parallel to the ground. More importantly, it means that the dipole will operate most efficiently when used with other horizontally polarized antennas. The dipole is discussed here because it is one of the antennas used in antenna specifications to describe antenna gain.

A simple base station antenna is the ground plane, shown in Fig.

Fig. 4-4 Ground plane antenna. (*Courtesy*, Avanti Research and Development, Inc.)

4-4. It receives its name from the metal elements which radiate from its base. These elements are called *radials*. Electrically, the ground plane operates as if it were a dipole mounted vertically. The radials act as an electrical substitute for the ground. A normal vertical antenna would be erected beginning at ground level. The ground, being a conducting medium, takes the place of one half of the dipole. In fact, an imaginary half dipole is created by the ground. Because an 11 meter signal is quickly absorbed by surrounding objects, it is desirable to raise the vertical radiator as high as possible. The radials will then act electrically the same way the ground does for standard verticals. The ground plane is an omnidirectional antenna, that is, it radiates energy equally in all directions. For this reason, it has less gain than the dipole.

The ground plane is the most common CB antenna. It is easy to use and does not require expensive devices to point it in the desired direction. It has a vertically polarized signal and communicates best with other vertically polarized antennas. Since virtually all mobile antennas are vertically polarized, vertical polarization is an advantage. The radiation from a ground plane is a low angle radiation. This is beneficial since signals which radiate at high angles do not contribute to communication.

Because low angle radiation represents a concentration of energy, the ground plane produces a signal gain over the isotropic radiator.

Several kinds of directional antennas exist. They all function by changing their radiation patterns to concentrate energy into a smaller space. The degree to which an antenna is successful in doing this is indicated by its *gain*. Gain is given as a comparison of the values of an antenna versus the calculated values of the isotropic antenna. If one antenna presents a gain of 3 dB over the isotropic antenna and a second antenna, a gain of 6 dB, the second antenna is said to have a more directional characteristic.

The abbreviations "dBi" and "dBd" are often used in antenna specifications. When an antenna is being compared with an isotropic radiator, the abbreviation dBi is used. When it is compared with a dipole, dBd is appropriate. When evaluating antennas rated according to these different standards, a correction factor of 2.14* must be used. If the antenna is compared to a dipole (gain is stated in dBd), 2.14 must be subtracted from the dBd gain figure. The resulting antenna gain figure will be comparable to antenna gain figures listed in dBi.

Antenna gain is important to the CB operator. Since the power of CB transmitters cannot legally be increased beyond a 4 watts output on AM or 12 watts output (PEP) on SSB (see Chap. 5 for discussion of output power), antenna gain is the only way of increasing the received signal strength. Antenna gain is combined with power output to provide effective radiated power (ERP). If the transmitter provides 4 watts output and six-tenths of a watt is lost because of the transmission line and matching losses, the antenna would radiate 3.4 watts. If the antenna had a gain of 3 dBi, the effective radiated power would be approximately 6.8 watts since 3

TABLE 4-1   Power Ratios

dB	Power ratio	dB	Power ratio
1	0	9	8.0
2	1.575	10	10.0
3	2.0	11	12.6
4	2.5	12	16.0
5	3.15	13	20.0
6	4.0	14	25.2
7	5.0	15	32.0
8	6.3		

dB equals a power ratio of 2. In Table 4-1, notice that each increase of 3 dB represents a doubling of power. With 3-dB antenna gain, the ERP of the above system is 6.8 watts. With 6-dBi gain, the ERP increases to 13.6

*The gain of a dipole over an isotropic is 2.14 dB.

watts. With 12 dBi gain, the ERP would be 54.4 watts. This increase in power is achieved by concentrating the antenna radiation into progressively smaller areas.

As indicated above, there are several types of directional antennas, most of them operating on similar principles. The directional antenna is composed of elements whose lengths are determined by the frequency for which they are designed. A common type is the beam. One element of the beam connects to the transmission line so that signals can be delivered to and received from it. This element is called the *driven element.* The driven element has a length equal to the wavelength for the frequency in use. Often some type of matching or tuning device will be provided so that a maximum amount of energy can be transferred to the antenna. Another element is longer than the driven element. Its length is determined by the distance of its separation from the driven element. The purpose of this longer element is to reflect the driven element's energy back to the front of the beam and to shield the driven element from signals which come from stations to the rear of the beam. This element is called a *reflector.* Shorter elements, called *directors*, are placed at the front of the beam. Their length is also determined by their spacing from the driven element. Directors concentrate the driven element's energy into a smaller pattern. The directional actions of both reflectors and directors function during reception as well as transmission.

A horizontal beam is simply a dipole with reflectors and directors added. Up to a point, more elements mean more gain. However, more elements require greater beam length and consequent mounting difficulties. A horizontal beam, like the dipole, is horizontally polarized. For this reason, they are not commonly used by CB operators.

A vertical beam is essentially a horizontal beam mounted in a vertical plane (see Fig. 4-5). These beams are very popular with CBers. Because a lot of CB communications takes place between base and mobile stations, vertical polarization has a definite advantage. Vertical beams are constructed similarly to horizontal beams. Reflectors and directors flank the driven element. Gain ranges all the way to 18 dBi and more. Beam length on some of these behemoths is over forty feet.

A third type of beam is available. It is something of a hybrid between the horizontal and vertical beams. Since it includes elements in both planes, dual polarization results (see Fig. 4-6). There are times when the characteristics of one type of polarization have advantages over those of the other type. With dual polarity, the benefits of both horizontal and vertical radiation can be tapped.

Phased Antennas

The antenna shown in Fig. 4-7 is an interesting type of directional antenna. It is composed of three vertical dipoles set in a wavelength-related pattern of distance. A control at the transceiver location supplies

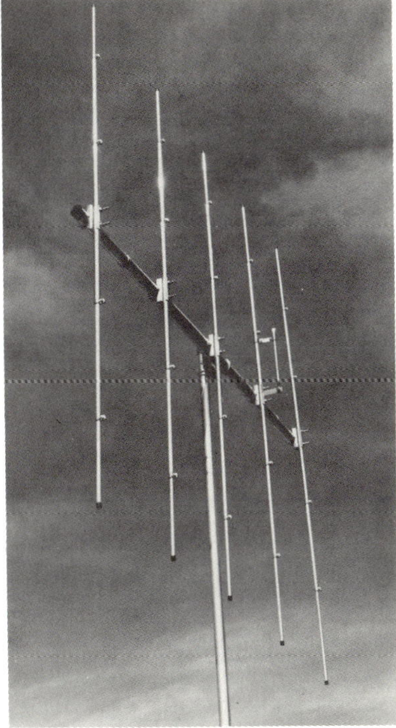

Fig. 4-5 Vertical beam antenna. (*Courtesy*, Cushcraft Corporation)

varying amounts of power to each of these dipoles. The antenna can be adjusted for omnidirectional operation at one time and directional operation at another. The real benefit of this antenna is that it does not require mechanical rotators. It can be used in the omnidirectional mode to establish a contact or to monitor the band, and it can be adjusted to improve communication by peaking it in the direction of the other station after communication is established.

Mobile Antennas

Mobile antennas are variations of some of those already discussed. Most of them operate on the ground plane principle. The difference is that the metal of the car or truck forms the ground plane portion of the antenna. This is convenient since the metal is available, but it has its problems in that the antenna radiation pattern is determined by the location of the antenna on the vehicle.

The best mobile radiator is the quarter wavelength whip. The problem with this antenna is one of height. Since it will project approximately 108 inches above the automobile, mounting it anywhere besides the bumper becomes a problem. If it is raised to the trunk level or to the top of

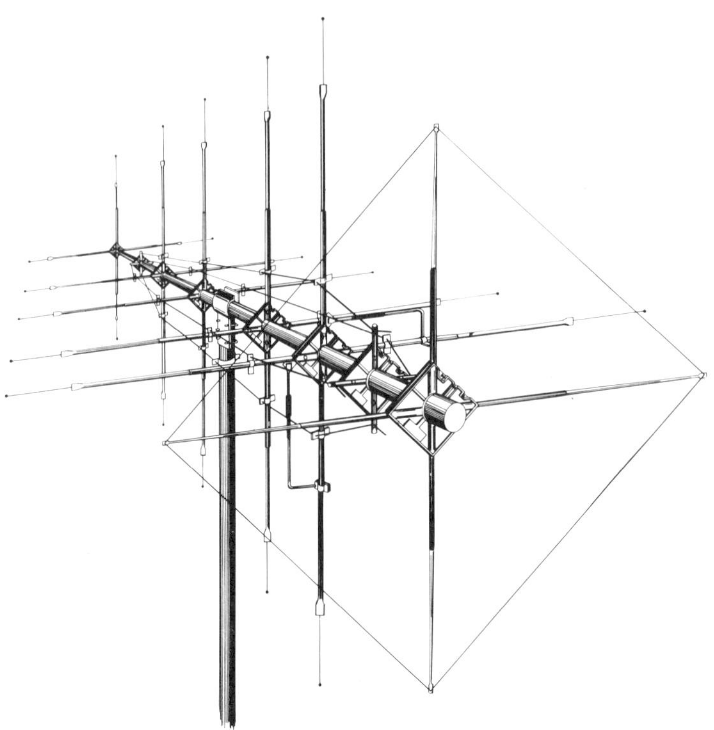

Fig. 4-6 Dual polarity beam antenna. (*Courtesy*, Avanti Research and Development, Inc.)

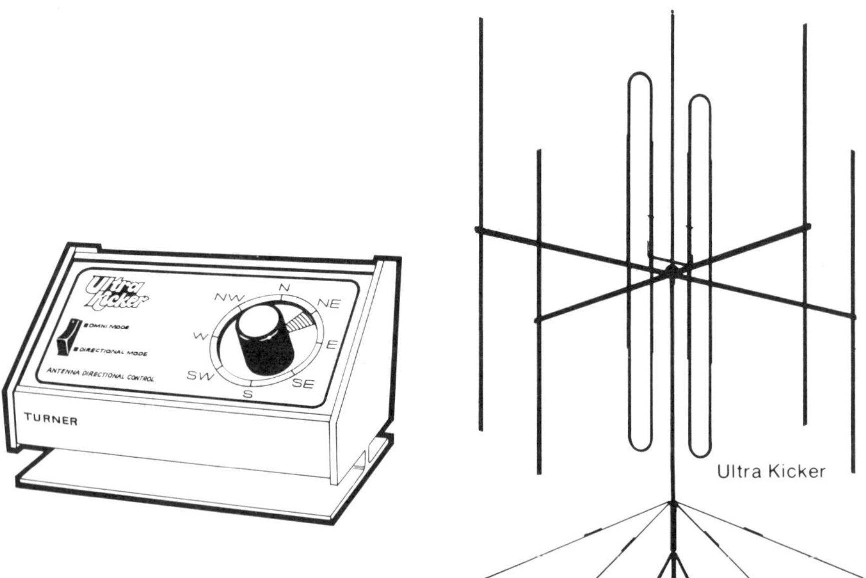

Fig. 4-7 Phased beam antenna with control unit. (*Courtesy*, Turner Division of Conrac Corporation)

the vehicle, overpasses and stop lights may be within its reach. Mounting the whip at bumper height has its problems as well. The lower position limits its distance. The bulk of the vehicle is now in front of the antenna, and since the pattern is not circular, the unit will transmit and receive better in some directions than others. Directional operation is a liability, not an asset, for mobile stations. (Figure 3-12 illustrated the changes in the mobile antenna radiation pattern which are caused by the mounting location of the antenna. As shown, the preferred circular pattern is produced by an antenna placed at the center of a vehicle.)

Short mobile antennas are popular. Short antennas use coils to reduce their length while insuring that the transmitter will be able to achieve a match with them. The lengths of these coil-loaded antennas range from 20 to 80 inches. The longer the antenna, of course, the better it will operate. After saying that, however, it must also be said that a shorter antenna mounted higher on a vehicle may perform equally as well or better than a longer antenna mounted lower on the vehicle.

Three basic types of coil-loaded antennas are in common use. Figure 4-8A shows a typical base-loaded antenna. The base-loaded antenna has a rigid coil assembly mounted at its bottom. A short whip screws into this coil assembly. The result is a sturdy, fairly efficient antenna of pleasing appearance.

A center-loaded whip (shown in Fig. 4-8B) is an improvement over the base-loaded design. Since the coil is placed at the center of the whip, the efficiency is improved. Placing the coil at the top of the whip would improve the radiation characteristics even more, but the rigidity involved and other considerations dictate against most top-loaded mobile antennas. Figure 4-8C shows a popular top-loaded antenna which has overcome these drawbacks.

Fiberglas antennas have become popular. These antennas either utilize helical winding throughout their length or enclose a loading coil at their top. The fiberglas antenna shown in Fig. 4-8D is a short, high-performance antenna which is durable and attractive.

The center-loaded antenna is very popular with truckers. They are often mounted on the side mirror brackets as shown in Fig. 4-9. The coil is above eye level, thus reducing the distraction which might result if it were lower. Also, the higher point of radiation increases antenna efficiency.

Radiation pattern distortion caused by the location of a mobile antenna can be reduced in one of several ways. The ideal location is at the center of the vehicle. For a standard automobile this would be the center of the roof. The second best position would be at the center of the vehicle at the top of the trunk lid. A single antenna mounted anywhere else on a car will display a directional pattern. One way of improving the pattern and increasing signal strength is to use two antennas mounted in such a way as to counteract the nonsymmetrical ground created by the metal of

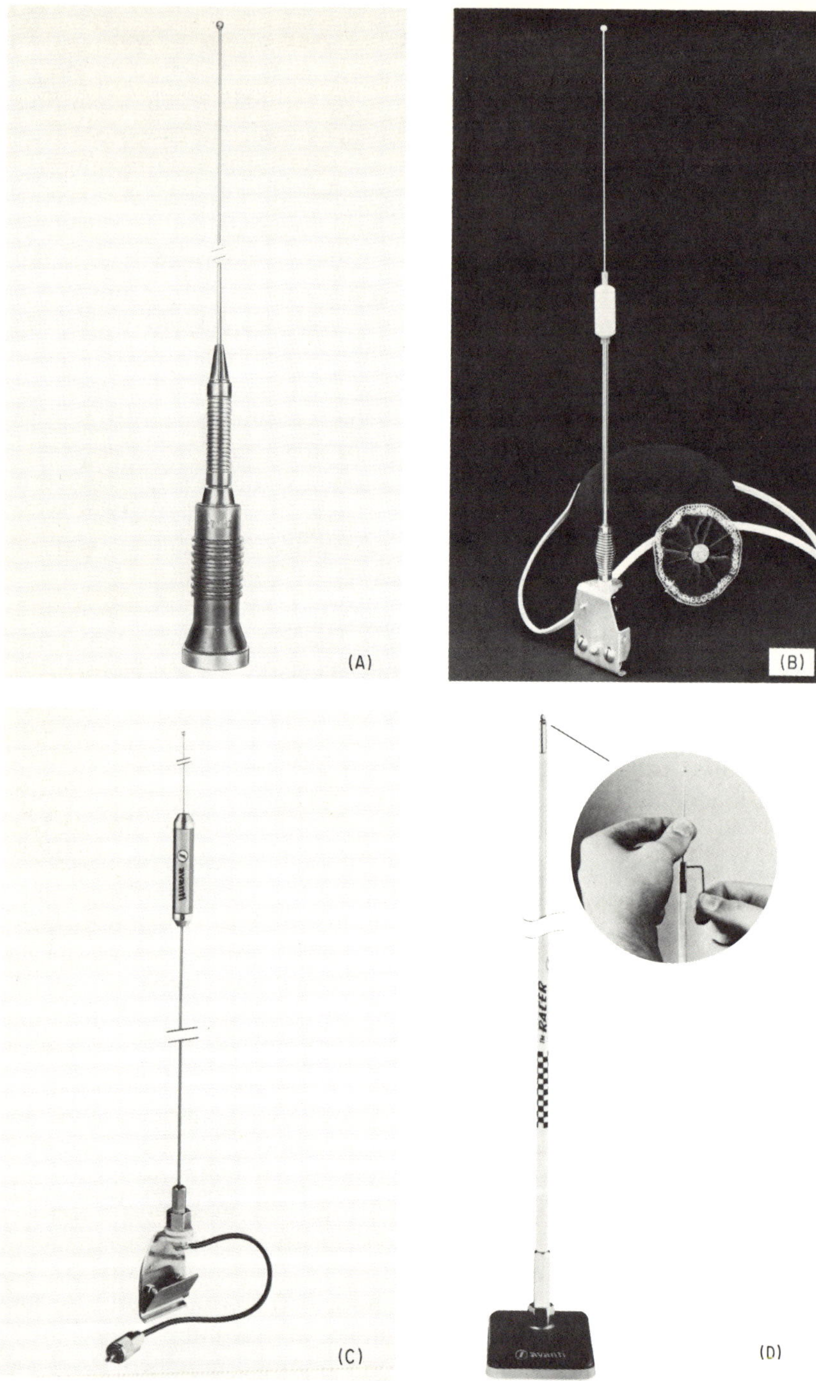

Fig. 4-8 *Mobile CB antenna:* (A) Base-loaded; (B) center-loaded; (C) top-loaded; (D) fiberglass. [(A), (C), and (D) *Courtesy*, Avanti Research and Development, Inc.; (B) *Courtesy*, Breaker Corporation]

Fig. 4-9 Co-phased center-loaded whip. (*Courtesy*, Breaker Corporation)

the vehicle. Two bumper-mounted whips could be used, one at each side of the rear of the car. Two loaded whips can be paired on either side of the vehicle. Trucks often use a pair of center-loaded whips attached to the mirrors. Less aesthetically pleasing is the practice of placing one antenna at the rear of the vehicle on one side and a second at the front on the other side. Looks may suffer, but the pattern shapes up nicely. With all dual antenna systems, a special cable harness is required to insure that percentages and phasing of power between the two antennas are correct.

When selecting an antenna, the CBer must consider its characteristics and intended use. His choice should be based on theory, preference, budget and experience, not necessarily in that order. For the conservative bank president, the smaller antenna may be the better choice even though the range is less. For the teenager seeking to impress the girls, an eight-foot whip on each corner of the car, all with fox tails flying may be just the thing. If the owner likes it and it works, then it is the best antenna around.

Installation of Fixed Station Antennas

The FCC has issued regulations concerning the installation of base station antennas. These regulations deal primarily with height. A distinction is made between omnidirectional and directional antennas. An omnidirectional antenna is one which has a circular pattern as seen from above the antenna looking down. The official definition describes an omnidirectional antenna as "an antenna designed so the maximum radiation in any horizontal direction is within 3 dB of the minimum radiation in any horizontal direction." The result is a relatively round pattern. Antennas which conform to this definition are omnidirectional antennas; those which do not are by definition directional. The importance of this distinction is that directional antennas are more limited in their height than are omnidirectional antennas. Directional antennas must be mounted so that their tops do not project more than 20 feet above the ground or above any man-made object or natural structure (tree, cliff, etc.). A man-made structure is anything made by man (house, barn, storage tank, etc.) except masts, towers, or poles. Figure 4-10 shows several different antenna

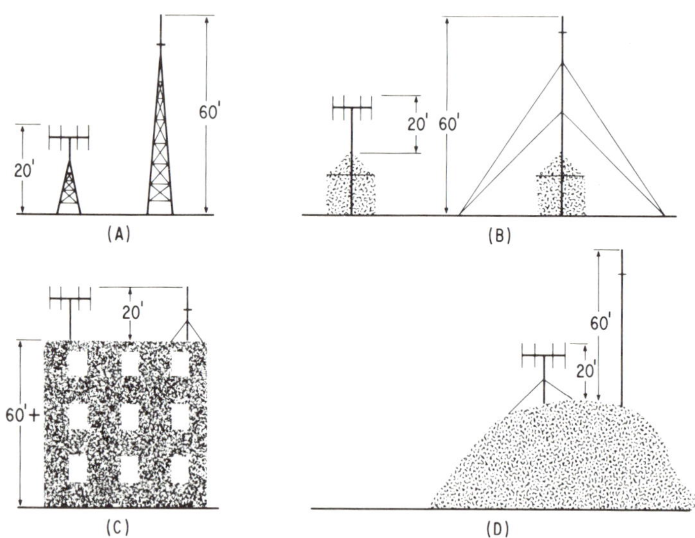

Fig. 4-10 *Antenna height regulations:* (A) Self-supported antenna structure—directional, 20' from ground; omnidirectional, 60'; (B) antenna support attached to man-made structure—directional, 20' above structure; omnidirectional, 60' above ground; (C) atop structure over 60' tall—20' above structure; (D) atop natural formation—directional, 20' above formation; omnidirectional, 60' above formation.

installations. Notice that the support structure determines what the permitted height of the antenna will be.

Omnidirectional antennas are permitted a height of 60 feet above ground level. This measurement is also referenced to the top of the antenna. It should be noted that the omnidirectional antenna height is stated as being measured from the ground, not from man-made objects. If a person possesses a valid license for some other type of radio station which is permitted antenna structures of greater height, a CB antenna, directional or nondirectional, can be mounted on the antenna support structure of the other station at a height not greater than 60 feet above the ground. Again, the measure is from the ground to the top of the antenna. The antenna cannot project above the support structure (tower or mast) of the other station even if this is less than 60 feet. One last height regulation is imposed on CB antennas. The top of the antenna must not "exceed one foot in height above the established airport elevation for each 100 feet of horizontal distance from the nearest point of the nearest airport runway." Unless a station location is less than a mile from an airport, this restriction should be of little concern.

Once the type of antenna to be employed has been chosen, the maximum height of the antenna is thus already known (60 feet for omnidirectional antennas and 20 feet for directional antennas). The next requirement is for a way to mount the antenna. The simplest mounting device is a pole or pipe. The antenna can be mounted on top of the pole, and the pole can then be raised and anchored in place. If the pole is long or the antenna heavy, this simple type of mounting can suddenly become complicated and even costly. Raising a long pipe with an antenna mounted on top is tricky, to say the least. The pole will tend to bend and perhaps break because of the stresses to which it is subjected. Its bottom may raise off the ground as the center of gravity is passed by the erectors. If it does, there may be dangerous problems. A pole unanchored is very hard to handle, and the antenna may come crashing down.

The most important thing in this kind of installation is preplanning. It is very important to survey the proposed antenna location to be sure that no electrical wires are close enough to pose a threat. The antenna installer should always plan his work with regard for Murphy's Law, which says: "Anything which can happen will happen and at the least opportune time."

Three accessories can be used to make the job easier and safer. Some sort of bottom stabilizer is needed to hold the bottom of the pole on the ground and in place. A second useful tool is a temporary stabilizer. This need be nothing more than a 2 x 4 plank (see Fig. 4-11). The stabilizer is placed behind the pipe and temporarily attached to it. The third accessory is a gin pole. Again, a 2 x 4 will serve adequately well. The gin pole has a pulley at its top. The erection rope passes through this pulley and attaches to the top of the stabilizer. When the rope is pulled, the pole is lifted from

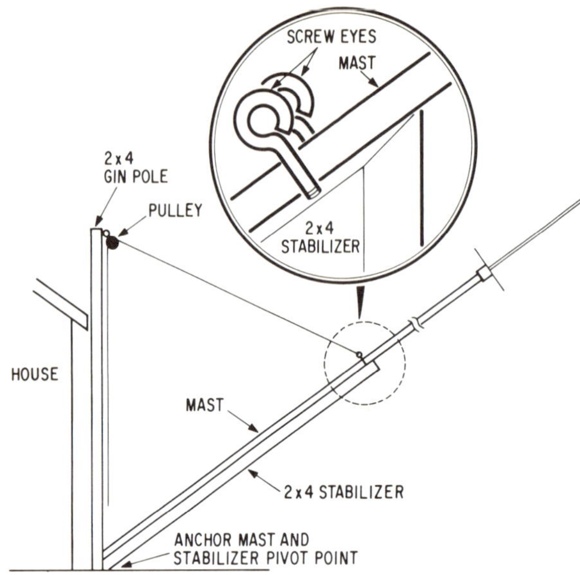

Fig. 4-11 Erecting an antenna.

a point above its center of gravity, thereby greatly reducing the stress. It must be remembered that the top of the gin pole may be beyond reach when the antenna is installed. The system must be designed so that the rope and temporary stabilizer can be easily released.

An easier support to erect is the telescoping mast. This device is constructed of several pipes, each just enough larger than the next so that the larger will permit the smaller to slide inside it. The masts are usually crimped and flared so that one pipe cannot be pulled entirely out of another. This is a very desirable feature. If a pipe were pulled out of the lower mast section during erection, it would undoubtedly destroy the antenna and possibly injure the installer.

Other features of the telescoping mast are the cotter pin stop and the set screw holder. When a mast section is fully extended, a cotter pin can be inserted through a hole in the lower mast section below the bottom of the extended section so that the top section cannot slip back into the lower section. The set screw is tightened to prevent the antenna from rotating.

It is important when erecting a telescoping mast to be sure that the bottom section is as plumb (straight up and down) as possible. If it is, the mast will endure very little strain during its extension since the weight will be pushing straight down. If the bottom section is installed at an angle, however, the stress may surpass the breaking point of the mast before it is fully extended.

With any mast of sizeable height, it is advisable to use guys for stability and security. The necessary number of guys depends on several things: the total height of the antenna structure, the weight of the antenna and its wind resistance, and the location of the bottom support. As a general rule, a set of guys should be placed as close under the antenna as possible without actually interfering with it. Caution must be used to insure that the bottom elements of a vertical beam will not snag the guys when they are rotated. Other guys should be placed at intervals no greater than 20 feet and preferably less. Figure 4-12 shows an antenna atop a 40-foot mast. The mast is connected to the gable end of a house at 16 feet. A set of guys is placed under the antenna at 40 feet. There are 24 feet between the gable mount and the top guys. A second set of guys at 28 feet will provide good stability for the mast. If a heavy beam antenna and rotor

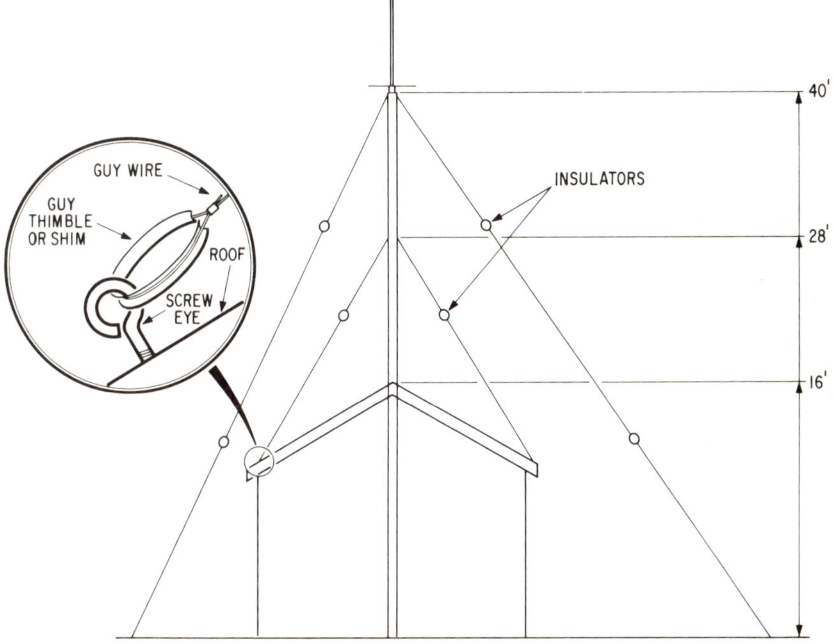

Fig. 4-12 Guying an antenna.

are used on this mast, or if the vicinity is subject to high winds, it would be advantageous to use two sets of center guys. The added stress caused by the heavier antenna would then be distributed between more guys.

It is a good practice to use insulators in guy wires. The reason is that lightning will be less likely to travel down the guys to the roof of the house, and unwanted resonances of the guy lines will be precluded.* Egg

---

* A resonant guy wire is one whose length matches the wavelength of the signal being transmitted. Resonant guy wires can alter radiation patterns and reduce signal strength. (See Chap. 4 for a discussion of resonance.)

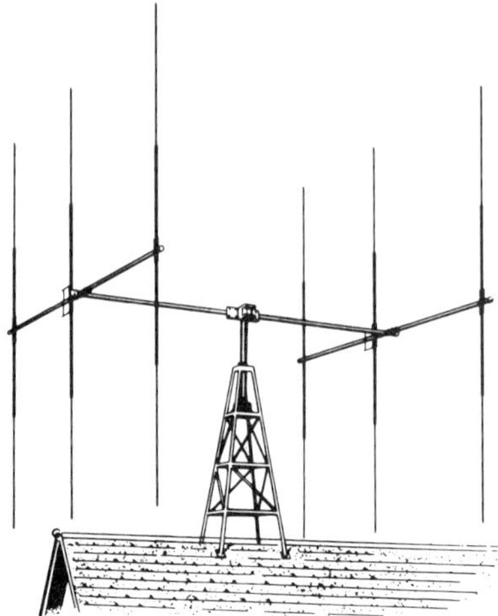

Fig. 4-13 Roof-mounted tower.

insulators should be used for this purpose. The construction of the egg insulator insures that the integrity of the guys will be retained even if the insulators should break. The wires pass through opposite ends of the egg insulator in such a way that the loop of one guy wire passes through the loop of the second guy wire. If the insulator is broken, the wires will continue to hold. The mast itself should be well grounded. Grounding helps to dissipate electrical charges and will better protect the house in case of a lightning hit.

During the erection of an antenna structure, it may be difficult to prevent the guys from tangling. Kinks in the guy wires must also be avoided. Kinks weaken the wire and can cause it to break. One method of reducing these problems is to place each guy through its screw eye prior to raising the mast. As the mast is extended, the guys will be played out without tangling and kinking. Even so, it is best to have one person at each guy point in case of kinks or hangups. Raising a tall antenna should be a neighborhood affair. If the operator will make a large pitcher of tea and invite the neighbors over, everyone will have a good time and serious mistakes will be less likely.

Other types of antenna support structures are available. A roof-mounted tower (see Fig. 4-13) is useful for CB antennas. Added to the length of the antenna, such supports can easily supply the legal height limit for directional antennas. The roof-mounted support is usually of a tripod design. Attachment is by lag screws through special feet on the support legs into the roof. Seals for the lag screw holes are normally provided with the tower to prevent leaks. Guying of roof-mounted towers may or may not be required. A lot depends on the antenna size, the tower height, and the roof construction. The CBer will have to analyze his own situation and respond accordingly. The best rule of thumb is, "If in doubt, guy it."

Full-length towers enjoy a great deal of popularity. They are more sturdy and can carry greater loads than other supports. If antenna maintenance is required, many full-length towers permit a person to climb up to the antenna instead of having to bring it down as would be the case with a telescoping mast. Depending on the height and windload, guying may or may not be required. Again, the CBer must decide in accordance with his situation. The manufacturer can usually provide assistance in choosing the proper tower for the job. Antenna weight is one consideration, but equally important is the amount of wind loading which the antenna, coax, and peripheral equipment create. Towers are designed to withstand a certain wind load at a certain height. The tower should be carefully chosen to accommodate the antenna that the CB operator intends to use.

A full-length tower requires a good anchor at its base. For shorter towers with lighter loads, a special base can be obtained. The base is designed so that the tower will mount to a metal plate. The plate has a metal pipe or stake attached to it which is driven into the ground to secure the bottom of the tower. Taller towers with heavier antenna loads may require a concrete mount. The size of this mount will vary according to the tower. The manufacturer is the best source of information for the proper base size and type.

When erecting a full-length tower, the gin pole is often useful. The gin pole is attached to the top of each tower section as it is erected. The rope is then attached to the next section slightly above its center. This section can now be raised to a position above the last section, turned until the sections mate, and then lowered into place. Using a gin pole, the process is easy and safe. Without it, tower erection is dangerous at the very least and may be impossible.

As indicated earlier, guying of the tower may or may not be required. It is often possible to brace the tower against a house, barn, or other type of building. If the tower does not extend far beyond this bracing point, guying will probably not be required. If the tower is free-standing or extends far above the building roof, guying probably will be necessary. Besides reducing the need for guying, using a tall building as a brace has another benefit. If the tower is connected to the building, the antenna can extend 20 feet above the building. If it is not, directional antennas must be limited to 20 feet above the ground. Omnidirectional antennas can be only 60 feet above ground in either case.

Three other types of towers are available. They are more expensive than the types mentioned above, but they may solve problems which will justify their increased cost. The self-supporting tower (Fig. 4-14) is useful

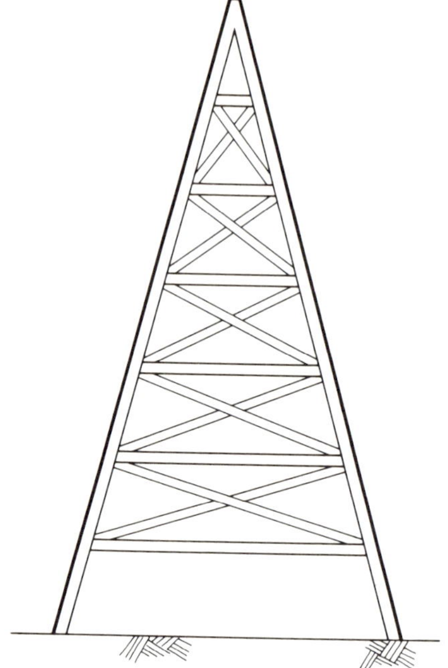

Fig. 4-14 Self-supporting tower.

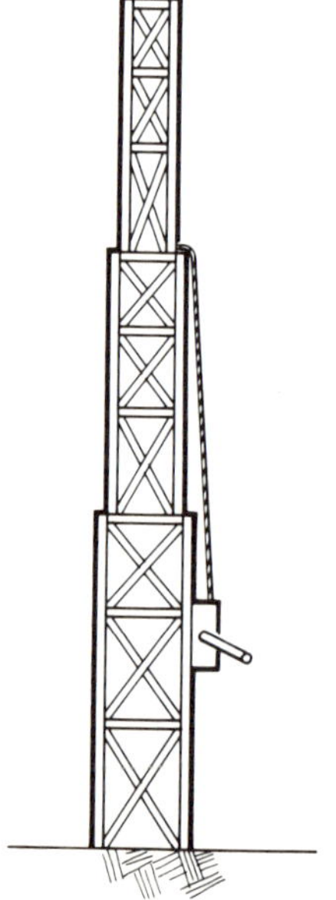

Fig. 4-15 Crank-up tower.

if the available space is small. It is designed to support itself and its load without external strengthening or guys. Instead of a guy field several yards in diameter, all that is required is room for a tower base, which will probably be only a few feet in diameter. It must be pointed out that in addition to the increased cost of the tower itself, the cost of the necessary set of tower anchors will also be high. Depending on the size of the tower and its load, many pounds of cement may be required to stabilize it.

A crank-up tower is another convenient type of tower (see Fig. 4-15). It consists of several sections, one inside the other. The sections telescope together, and when they are all nested, the tower height is only slightly greater than that of a single section. A crank and cable permit the tower to be extended while the operator remains on the ground. If antenna maintenance is required, the tower can be cranked down to its minimum height and the work done at a much safer and more convenient height.

The fold-over tower (Fig. 4-16) provides a convenient method of antenna maintenance. The tower is constructed of two parts. The base section is simply a sturdy, free-standing tower. The fold-over section hinges at the top of the base section. A crank reels in and reels out a cable

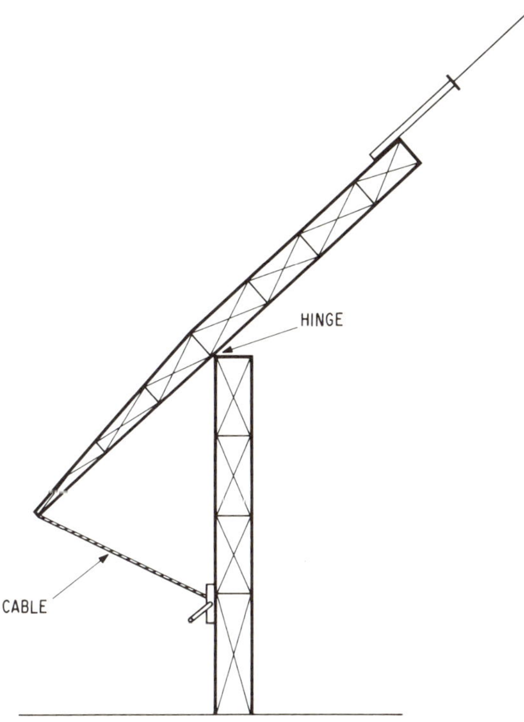

Fig. 4-16 Fold-over tower.

that is connected to the bottom of the fold-over section. When antenna installation or maintenance is required, the cable is cranked out until the top of the fold-over section can be reached from the ground. When the antenna work is completed, the cable is reeled in, bringing the fold-over section back to the vertical.

Attachment of antenna support structures to buildings and houses can take many forms. The saddle mount is a very useful mounting device. Since most houses are constructed with eaves jutting out beyond the walls, it is not possible for a mast to be attached directly to a wall. The saddle mount attaches to the wall and projects outward so that the mast can be fastened to it. Saddle brackets can be obtained in several lengths of projection, ranging from 8 to 16 inches. In construction, they are simply a strip of heavy metal which has been drilled and bent as shown in Fig. 4-17A. Two lag screws or masonry anchors should be used to attach the bracket to the wall or eave of the building. A horseshoe clamp is then used to hold the mast into the saddle of the bracket. The result is a sturdy and attractive means of attachment.

Mounting a mast on the peak of the roof requires a special device. The peak-mounting bracket has a receptacle for the bottom of the mast and a couple of mounting plates which rest on either side of the roof. Special care should be taken to prevent the mast or support from breaking or puncturing the ridge row shingles. Also, sealers should be used to guard against leaks where the screws enter the roof. A peak-mounting bracket requires that the mast be guyed.

Chimneys and roof pipes are often convenient for attaching masts. A mast may be attached to a chimney by either of two methods. Side-mounting brackets can be attached to the chimney with masonry anchors and the mast secured to the saddle of the bracket with a horseshoe clamp. Chimney straps can be used that do not require drilling holes into the chimney. These straps wrap around the chimney and provide saddle mounts on one side (Fig. 4-17B). Horseshoe clamps are again used to hold the mast in place. Depending on the height of the chimney and the height of the antenna structure, guying may or may not be required.

If pipes project through the roof, they may be convenient attachment points. A bracket such as is shown in Fig. 4-17C can be used to attach the mast to the pipe. Horseshoe clamps are used on both the pipe and mast. If the pipe and mast diameters are the same, clamps can be used to hold the two together without other devices. Guys will usually be required when pipe mountings are used.

Guy anchors can be simple or elaborate The simplest anchor is a heavy screw eye, particularly useful for anchoring a guy to the roof of a building, preferably in a rafter. The screw eye should not be located too near the end of the rafter nor so far from the end that it might create a leak in the building.

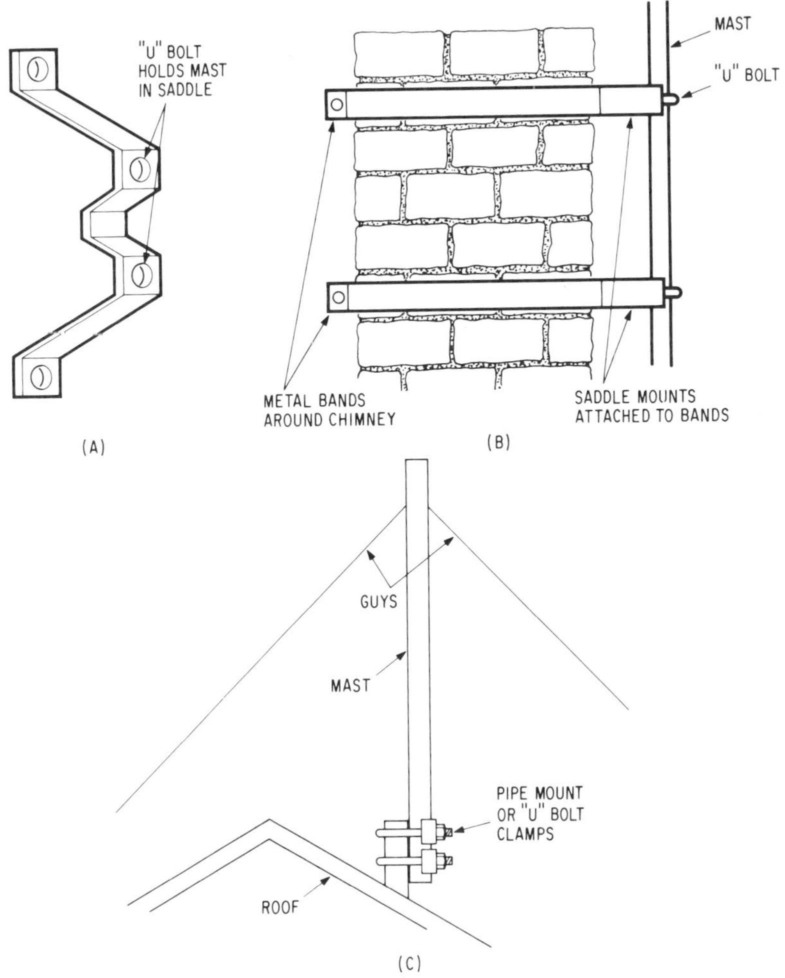

Fig. 4-17 *Mounting devices:* (A) Saddle mount; (B) chimney mount; (C) pipe mounting.

Guy shims are useful to prevent wearing of the guy wire under high winds. The shim is placed through the eye of the screw eye, and the guy wire is then laid in its channel and tied off (see inset of Fig. 4-12). The wear will now be on the guy shim and not on the wire itself.

Guying into masonry is more difficult but still possible. Lead split shell anchors are useful for this task. A proper sized hole is made in the masonry, and the shell is placed into it. A screw eye with lag bolt threads is then screwed in. In the process of screwing, the two halves of the shell spread, causing it to wedge in the hole. It is best to place the anchor so that the guy pulls obliquely to the hole and not straight outward.

Rotators

Directional antennas usually require the use of rotators. Rotators present problems of their own. First, they add weight to the top of the support structure. The strength of the mast and the amount of guying must be capable of securing the weight of the antenna and the rotator. The rotator must be powerful enough not only to support but to turn the antenna load. If two beams, each with forty-foot booms, are stacked side by side, something more than a standard television antenna rotator will be needed. For this purpose, heavy-duty rotators are available. These are the rotators used by Amateur Radio Operators for their large beams. Their cost is greater than that for television antenna rotators, but in the long run, the added expense should prove its worth.

Even heavy-duty rotators may be overburdened by the weight of some antennas. A thrust bearing can assist in reducing the rotator load. The thrust bearing, mounted between the rotator and the load, supports the weight of the antenna, reduces the wear on the rotator, and makes the turning of large antennas easier.

Rotators mount in several different ways. A rotator for small antennas may mount at the top of the mast. A heavy-duty rotator will mount lower and may have a thrust bearing above it. Some rotators mount in towers. Thrust bearings may or may not be used above them.

Rotators require control cables to be run up the mast. The antenna requires coax. For greatest durability and attractiveness, these cables must be attached to the antenna support with standoffs. Some standoffs will accept both cables. Other standoffs will require separate runs for each cable. Whichever method is used, the installation should be neat and durable.

CHAPTER 5

———

Improving CB Communications

Communications on the Citizens Band are often marginal. Propagation for band changes and signals from far away may be heard as strong as locals. Problems can occur at any time simply because the number of users has become too great. Interference from stations on adjacent channels may decrease the reliability of contact. At such times, CBers seek methods of improving CB communications. This improvement can be one of two kinds, that gained by operating methods and that gained by technical means.

Operating Methods for Improved Communications

Standard operating practices were discussed in Chaps. 2 and 3. This section describes more specific operating techniques. Many of these have evolved from the requirements of everyday CB activity. This situation is both good and bad. On the one hand, techniques developed and refined in actual practice are more likely to fit the unique needs of the CBer. On the other hand, procedures developed in this way may lack the standardization which promotes improved communication. Some CB groups are working for the standardization of various procedures and techniques. Their success will be decided by the cooperation or lack of cooperation on the part of individual CB operators.

Calling Channels

At one time, channel 11 was designated by the FCC as a national calling frequency. Stations would make initial contact on channel 11 and would then switch to an alternate channel to carry on their conversations. The benefit of this procedure was that stations desiring to establish contact did not have to wait for lulls in lengthy conversations to call one another. Also, they could arrange beforehand to meet on channel 11 with greater assurance that they would find the channel clear. Due to the crowding of the CB band, the FCC decided that this channel could be used to greater advantage if it were an open rather than a calling channel. The FCC therefore rescinded the order making channel 11 an official calling frequency.

Unofficial calling channels have often been established. These channels differ in different areas. Where they exist, CBers voluntarily use the designated channel for calling only. When such calling channels are respected and used properly, all CBers benefit.

Breaking

If no calling channel is available, a different procedure is helpful. When one station is transmitting on a channel, it is inaccessible to other stations in the area. If another station has traffic (communications) to share, he will attempt to break into the conversation that is in progress. The CBer seeking to interrupt another conversation is called a *breaker*. To let the other stations know of his need to use the channel, he will wait until one party is completing a transmission and will then call for a break. He will say "break, break," or will follow the word "break" with a channel number, for example, "break, nineteen," or "break, one-nine" Both of these breaks are for channel 19. A more proper break format uses the word "break" followed by the breaking station's call sign. This break takes more time, however, and the simple breaks given above are more common.

It is important for stations to respect the rights of breakers. The Golden Rule of CB radio is, "Break for others as you would want them to break for you." By the same token, breakers should not abuse their privilege, either. If a person needs to meet a schedule with another operator on a particular channel which is in use, he may establish contact by breaking. Once contact is made, the breakers should move to another channel. If a CBer is seeking information from local stations, he should break into a conversation, obtain his information, and then withdraw from the channel unless requested to join in the conversation by the stations using the channel. When breaking, it must be remembered that a station cannot hear while it is transmitting. Breaking must occur at the end of a transmission so that the receiving station can hear the breaker and respond when he begins his transmission.

Transmission Length Limitations

The FCC has established rules concerning the length of CB conversations. To insure that all stations have an equal chance to use the CB frequencies, interstation communications must be limited to five minutes. Following five minutes of interchange, a one-minute silent period must be observed. If communications are still desired between the stations, they can resume after the one-minute silent period providing that the band is not then in use. It is prohibited for the stations to change frequencies to permit continuation of communications before the one-minute silent period has elapsed. Communications between units of the same station are not limited by the five-minute rule. However, units of the same station are expected to transmit only as long as required to complete the interchange. In other words, communication should be completed as soon as possible.

Communication Words and Codes

When band conditions are poor and when the length of communications must be kept short, standard words and abbreviations can be used to clarify and speed up transmission. Some words are in common use by all radio communication services. Table 5-1 lists these words with their meanings.

Two basic codes are used by CBers, the "10 signals" and the "Q signals." The "10 signals" use the number 10 as an identifier and a second number to specify what the communication is. Perhaps the two most common 10 signals are 10-4 and 10-20. 10-4 is used to acknowledge the receipt of a message with which one is in agreement. An example of its use would be for a dispatcher to transmit information and location of a service call to a service man. If the service man received and understood the transmission, he would reply very simply, "10-4." The signal 10-20 refers to the location of the unit. It is used both as a question and as a reply. For example, "What is your 10-20?" This means, "Where are you located?" The reply would be, "My 10-20 is at the corner of 4th and Brazos." The rest of the 10 signals are used similarly. Many can be used as both questions and answers. Once a person begins to operate on the Citizens Band, he will learn the popular 10 signals quickly. A list of those adopted by the Associated Public Safety Communication Officers, Inc. (APCO), is given in Table 5-2. Other lists of 10 signals with different

TABLE 5-1   Radio-Telephony "Shorthand"

Clear—Indicates that the conversation is concluded and that no further response is expected.

Out—An alternative to the use of clear; both mean the same thing.

Over—"My transmission is ended, and I expect a response from you." "Over" is used at the end of a transmission to indicate that the sendor has completed his part of the conversation and now wishes the other party to transmit.

Repeat—Means just what it says: "Say it again. I did not understand it the first time."

Roger—"I have received your transmission and understand it clearly." "Roger" is an acknowledgement that the transmission was received and understood.

Wilco—"I have received and understood your transmission and will comply with it." "Wilco" adds to "Roger" the idea of acting upon the directive of the message.

Words Twice—Words twice is requested when the information is very important and a high degree of accuracy is required or when the received quality is poor. The request is for each word or phrase to be repeated two times.

TABLE 5-2   Official APCO 10-Code

Code	Meaning	Code	Meaning
*10-1	Unable to copy—change location	10-49	Traffic light out
		*10-50	Accident—F, PI, PD
*10-2	Signals good	*10-51	Wrecker needed
*10-3	Stop transmitting	*10-52	Ambulance needed
*10-4	Acknowledgment	10-53	Road blocked
*10-5	Relay	10-54	Livestock on highway
*10-6	Busy—stand by unless urgent	*10-55	Intoxicated driver
*10-7	Out of service (give location afd/or telephone number)	10-56	Intoxicated pedestrian
		10-57	Hit and run—F, PI, PD
*10-8	In service	10-58	Direct traffic
*10-9	Repeat	*10-59	Convoy or escort
10-10	Fight in progress	10-60	Squad in vicinity
10-11	Dog case	10-61	Personnel in area
*10-12	Stand by (stop)	*10-62	Reply to message
*10-13	Weather and road report	*10-63	Prepare to make written copy
10-14	Report of prowler	*10-64	Message for local delivery only
10-15	Civil disturbance	*10-65	Net message assignment
10-16	Domestic trouble	*10-66	Message cancellation
10-17	Meet complainant	*10-67	Clear to read net message
*10-18	Complete assignment quickly	*10-68	Dispatch information
*10-19	Return to . . .	*10-69	Message received
*10-20	Location	*10-70	Fire alarm
*10-21	Call . . . by telephone	10-71	Advise nature of fire (size, type, and contents of building)
*10-22	Disregard		
*10-23	Arrived at scene	10-72	Report progress on fire
*10-24	Assignment completed	10-73	Smoke report
*10-25	Report in person to (meet) . . .	*10-74	Negative
10-26	Detaining subject—expedite	*10-75	In contact with
10-27	Drivers license information	*10-76	En route
*10-28	Vehicle registration information	*10-77	ETA (Estimated Time of Arrival)
*10-29	Check records for wanted	10-78	Need assistance
*10-30	Illegal use of radio	10-79	Notify coroner
10-31	Crime in progress	10-80	
10-32	Man with gun	10-81	
*10-33	EMERGENCY	*10-82	Reserve lodging
10-34	Riot	10-83	
10-35	Major crime alert	10-84	If meeting . . . advise ETA
*10-36	Correct time	10-85	Will be late
10-37	Investigate suspicious vehicle	10-86	
10-38	Stopping suspicious vehicle (give station complete description before stopping)	*10-87	Pick up checks for distribution
		*10-88	Advise present telephone number of . . .
10-39	Urgent—use light and siren	10-89	
10-40	Silent run—no light or siren	10-90	Bank alarm
*10-41	Beginning tour of duty	10-91	Unnecessary use of radio
*10-42	Ending tour of duty	10-92	
*10-43	Information	10-93	Blockade
10-44	Request permission to leave patrol . . . for . . .	10-94	Drag racing
		10-95	
10-45	Animal carcass	10-96	Mental subject
10-46	Assist motorist	10-97	
10-47	Emergency road repairs needed	10-98	Prison or jail break
10-48	Traffic standard needs repairs	10-99	Records indicate wanted or stolen

Note: Signals marked by an asterisk are those in most use.

meanings attached to some of them are also available. Use of the APCO list is to be preferred in order to further the cause of standardization.

Another set of abbreviations in common use is the Q signals. Just as "10" was the identifier for 10 signals, "Q" is the identifier for Q signals. Two letters follow the Q to specify meaning. QTH conveys the same meaning as 10-20. An example of its use is the following interchange:

"What is your QTH?"
"My QTH is 4th and Brazos."

The Q signals, originated in 1912, consist of 50 distinct signals. The original signals dealt with Morse code transmission. Some modification of the original Q signals is reflected in the list given in Table 5-3, which has

TABLE 5-3　"Q" Signals

	Question	Reply
QRA	What is the name of your station?	The name of my station is . . .
QRB	How far approximately are you from my station?	The approximate distance is . . . miles.
QRD	Where are you bound and where are you from?	I am bound for . . . from . . .
QRE	What is your estimated time of arrival . . . (place)?	My estimated time of arrival at . . . (place) is . . . hours
QRF	Are you returning to . . . (place)?	I am returning to . . . (place), or, Return to . . . (place)
QRL	Are you busy?	I am busy (or I am busy with . . .). Please do not interfere.
QRM	Are you being interfered with?	I am being interfered with.
QRS	Shall I speak more slowly?	Speak more slowly.
QRT	Shall I stop transmitting?	Stop transmitting.
QRU	Have you anything for me?	I have nothing for you.
QRV	Are you ready?	I am ready.
QRW	Shall I inform . . . that you are calling him on Channel . . .?	Please inform . . . that I am calling him on Channel . . .
QRX	When will you call me again?	I will call you again at . . .
QRZ	Who is calling me?	You are being called by . . .
QTJ	What is your speed?	My speed is . . .
QTN	When did you leave . . . (place)?	I left . . . (place) at . . .

TABLE 5-3 "Q" Signals (cont'd)

QTR	What is the correct time?	The correct time is . . .
QTU	What are the hours during which your station is open?	My station is open from . . . to . . .
QTV	Shall I stand guard for you on Channel . . .?	Stand guard for me on Channel . . .
QTX	Will you keep your station open for further communication with me until further notice (or until . . .)?	I will keep my station open for further communication with you until further notice (or until . . .).
QUA	Have you news of . . . (call sign)?	Here is news of . . . (call sign).
QUD	Have you received the urgency signal sent by . . . (call sign of mobile station)?	I have received the urgency signal sent by . . . (call sign of mobile station).
QUF	Have you received the distress signal sent by . . . (call sign of mobile station)?	I have received the distress signal sent by . . . (call sign of the mobile station) at . . .
QUM	Is the distress traffic ended?	The distress traffic is ended.
QUO	Shall I search for . . .?	Please search for . . .
QUR	Have survivors . . . (1) received survival equipment? (2) been picked up? (3) been reached by rescue party?	Survivors . . . (1) are in possession of survival equipment (2) have been picked up (3) have been reached by rescue party
QUS	Have you sighted survivors or wreckage? If so, where?	Have sighted . . . at . . .
QUT	Is position of incident marked?	Position of incident is marked by . . .
QSI	....	I have been unable to break in on your transmission, or, Will you inform . . . (call sign) that I have been unable to break in on his transmission on Channel . . .?
QSL	Can you acknowledge receipt?	I am acknowledging receipt.
QSN	Did you hear me (or . . . call sign) on Channel . . .?	I did hear you (or . . . call sign) on Channel . . .
QSO	Can you communicate with . . .?	I can communicate with . . .
QSX	Will you listen . . . (call sign) on Channel . . .?	I am listening to . . . (call sign) on Channel . . .
QSY	Shall I change to another channel?	Change to Channel . . .

been altered for use with voice transmissions. As with 10 signals, both question and reply can be represented by Q signals. Some signals have gained other usage as well. "QSO" is often used as a noun to indicate a conversation over the radio—"I enjoyed the QSO." "QSL," which refers to acknowledging the receipt of a transmission, has been printed on post cards used to confirm a contact. These cards, called "QSL cards," are usually printed with the station call and other information on them. The list of Q signals in Table 5-3 represents those most often used by CBers. FCC rules require that a list of Q signals and 10 signals used by a station be kept with the station records. This is also true of all other codes and abbreviations.

When conditions are not favorable to communication, it is helpful to spell out words letter-by-letter, using phonetic equivalents. Thus, "Noel" is spelled phonetically: "November-Oscar-Echo-Lima." Ambiguity of letter sounds is reduced by using the standard list of phonetic words given in Table 5-4.

A recent phenomenon among CBers is the use of slang. Words and phrases have been coined in a sort of CB dialect. The original purpose may have been to disguise a conversation from parties listening in. Some suggest that slang got its primary impetus from truck drivers seeking to

TABLE 5-4   Phonetic Pronunciation Table

Letter	Identifying Word
A	Alfa
B	Bravo
C	Charlie
D	Delta
E	Echo
F	Foxtrot
G	Golf
H	Hotel
I	India
J	Juliet
K	Kilo
L	Lima
M	Mike
N	November
O	Oscar
P	Papa
Q	Quebec
R	Romeo
S	Sierra
T	Tango
U	Uniform
V	Victor
W	Whiskey
X	X-Ray
Y	Yankee
Z	Zulu

avoid highway patrol officers. CB slang has become so extensive that it cannot be included in any complete way in the present volume. Appendix A provides a brief list of common CB slang words. The benefit for communications which results from this dialect is questionable. Nevertheless, it is becoming ever more popular. The CB operator must learn some slang out of self-defense.

Auxiliary Services

The popularity of CB operation has been increased by the auxiliary services it provides. Many travelers have found CB radio to be an indispensable traveling companion. The importance of CB radio during emergency situations has already been discussed. Besides the security and peace of mind provided to the stranded motorist or the person in trouble, CB radio is a very practical help for people traveling in unfamiliar territory. Information concerning directions, routes, and highways can be obtained. The closest lodging and the best restaurant can be found without having to consult a local telephone book. It is even possible to order meals or arrange for motel rooms by communicating with businesses equipped with CB radio. When a CBer travels in unfamiliar places, some local CBer can become his counselor and guide. It is only appropriate for the CBer to return the courtesy by assisting visitors to his own neighborhood.

Another service CBers provide for one another is the relaying of messages. Because the range of CB radio is short, a person may desire to contact a particular station to impart some information only to find that he is beyond the range of the station. One answer is for the CBer to contact a station closer to the desired station and ask for assistance. This intermediate station can relay information to the distant station and then report any reply to the calling station. The only restriction is that the relay shall not extend more than 150 miles.

Some CB operators use a phone patch at their base station. A phone patch is a device which permits a radio to be used in conjunction with a telephone. A person can contact a station which has a phone patch, and the operator of that station can place a call for him. During the telephone conversation, the base station operator will switch between transmit and receive as the conversation requires. A special transformer is used to change the single set of wires which carries signals both ways on the telephone circuit to the divided transmit and receive circuits required by the CB radio. Since the radio does not transmit and receive at the same time, the operator turns the transmit on when the mobile CBer wants to talk and off when he wants to hear. The result is that a CBer can make phone calls from a unit in his car or boat or from a portable unit.

Specific installation methods for the phone patch will be explained by the manufacturer and the local telephone company. The phone patch can usually be attached directly to the phone wires coming into the house. In addition, the unit will connect to the speaker output and to the microphone or some alternate input of the radio. When the phone patch is in use, the base station operator will find it necessary to monitor the conversation.

Technical Enhancement of Communications

Citizens Band communication can be improved through techniques of operation, as just explained. In addition, improvement can be obtained by technical means. Since CB radios are limited in the amount of RF output power that can be delivered, it is important that this power be used efficiently. Single side band transmission, speech processing, and antenna matching all help make the most of the power available. Reducing the amount of noise received will allow the desired signal to be heard better. All these subjects will now be discussed.

Single Side Band

Most Citizens Band radios use amplitude modulation (AM). Modulation is the process of adding meaningful information to an electromagnetic (radio) wave. This electromagnetic wave is called a *carrier* because it carries the modulation (the meaningful information). The sound which the human ear hears depends on air to transport it from one person to another. Air molecules are heavy, and sound energy is quickly used up pushing them around. Therefore, sound has a very short range under normal conditions.

An electromagnetic carrier wave obeys the laws of high frequency radiation, that is, it does not depend on air to transport it but radiates through space in the same way that light does. (Light is itself a form of electromagnetic radiation.) Carrier waves can travel long distances between a transmitter and receiver. Since sound, despite its limited range, can ride "piggyback" on this carrier wave, it, too, is enabled to travel long distances. One method of placing the sound on the carrier wave is called *amplitude modulation*. Figure 5-1 shows how the carrier wave is modulated with the sound. Notice that the sound wave rides on both sides of the carrier. These two sound waves are called *side bands*. Each side band has all of the sound information on it. The lower side band is the mirror image of the upper side band. If the carrier wave was split into two equal parts, the sound could be recovered in its entirety from the upper or lower side band alone. To do so would represent an economy of energy and radio frequency spectrum since the same amount of information would be transmitted in half the space and with half the power. This is the philosophy behind single side band transmission.

In a single side band transmitter, a filter in the low-power portion of the unit strips off one of the side bands, and therefore the output portion

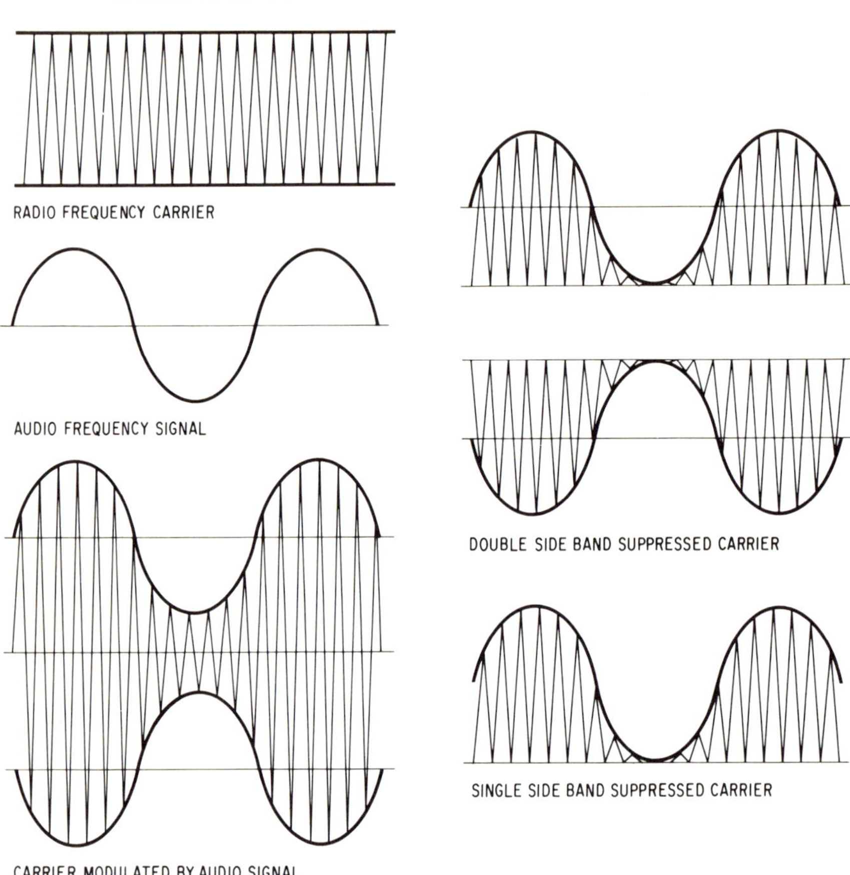

Fig. 5-1 Modulation.

puts all of its energy into only one side band, resulting in a higher percentage of meaningful information being transmitted.

A single side band transmitter increases the efficiency of the power in another way, also. Amplitude modulation of a carrier wave results in a signal that is 50 percent greater than the carrier wave alone. When the signal is received, the carrier and the sound are separated, and the carrier energy is discarded since it has no meaningful information in it. The sound energy alone is able to impart meaningful information, and it represents only one-third of the total transmitted power. This percentage is very inefficient. In amplitude modulation, when the sound wave modulation is equal to one-half of the carrier wave, it is referred to as one-hundred-percent modulation. If the modulation is increased beyond this point, distortion will result. If efficiency is to be increased, another method must be used.

Single side band, suppressed carrier is the most efficient method for transmission used in Citizens Band equipment. It turns the carrier on and off with the sound wave so that the louder the sound, the greater the power of the carrier. If no sound is present, no carrier is transmitted. The result is that one-hundred percent of the transmitted energy is meaningful information (sound).

Several aspects of single side band require clarification when it is applied to Citizens Band. First, the amount of power which is authorized is confusing. The Federal Communications Commission limits Citizens Band transmitters using amplitude modulation to a maximum power output of 4 watts. Single side band transmitters are permitted 12 watts output. On the surface, it appears that single side band is permitted much more power output than amplitude modulated units. This conclusion is not entirely correct. The measurement for AM units is an unmodulated measurement of the carrier power. At one-hundred percent modulation, an additional one-half of this amount is present. This amounts to 6 watts (4 watts carrier plus 2 watts sound). The single side band measure is a peak envelope power measurement, which means that it is measured at one-hundred percent modulation. Also, since this modulation is specified to be a sine wave tone, the signal is at a minimum value as often as it is at a maximum value. The actual power is therefore one-half of the total amount, or 6 watts (12 watts divided by 2 equals 6 watts). Do not let this confuse you about the superiority of SSB over AM. Even if the actual output power is equivalent, SSB units distribute their power over half as much frequency bandwidth, and one-hundred percent of it contributes to meaningful communication rather than the thirty-three percent of AM units.

Manufacturers who advertise units offering 120 channels are not validly reporting the actual number of channels available. This 120 channel figure includes 40 AM channels and 80 SSB channels. As described, SSB uses only one-half of an AM channel at a time. Therefore, in reality, there are 40 AM channels which can be utilized as 80 SSB channels. A channel can be used for one AM signal or two SSB signals. It is an either/or situation, either 40 AM or 80 SSB channels, but not 120 different channels.

Single side band has definite advantages over AM. Its usable range is greater, and twice as many channels are available.* Amateur radio operators learned quickly that SSB was superior to AM in range and that it provided at least a partial answer to the problem of band congestion. In all probability, Citizens Band radio operators will turn to SSB in increasing numbers as communications become ever more congested. The additional cost will be offset by increased performance.

---

* Simultaneous use of both side bands of a channel depends on the quality of the equipment.

Speech Processing

It was explained in the preceding section that the information conveyed over CB radio is equal to the percentage of modulation. If the transmitter is fully modulated (100 percent), the maximum amount of information will be conveyed. If the transmitter is modulated more than 100 percent, distortion results. If it is modulated less than 100 percent, the transmitter's "talk power" is reduced. CB operators therefore strive for the highest average modulation possible without distortion.

When a person talks with a hand-held microphone, he usually holds it close to his mouth so that the sound goes directly into it. Modulation can be set for a consistently high average, but if different people use the same microphone and the strength of their voices differ, one will modulate the transmitter well and the other will not. Since base station microphones are often not kept at a uniform distance from the operator's mouth, the level they produce will vary, and maximum modulation may not be maintained.

There are devices available which overcome these problems. Speech processors or speech compressors are amplifiers which automatically vary their gain. They operate by producing a uniform output level even though the input level may vary. A speech compressor amplifies soft and distant voices until the transmitter receives enough level for 100-percent modulation. If a person with a loud voice speaks over a standard amplifier adjusted for a soft voice, the amplifier will deliver too high a level, and distortion will result. To prevent this, the compressor automatically reduces the output so that the transmitter will not be overmodulated. When the loud voice ceases speaking, the compressor returns to the setting which is best for soft voices.

Figure 5-2 shows the output of a typical speech compressor relative to the input. The dashed line going up from the very soft voice position intersects the compressor curve at a point equal to 3 on the output level column. The dashed line from the medium voice position intersects the compressor curve at a level equal to 6 on the output level column. This is an increase of 3. The dashed line from the very loud position intersects the compressor curve at a level equal to 6.25 on the output level column. This is an increase of only one-fourth of one unit. The compressor action is apparent. From a soft voice to a very loud voice, the output level will not change over 1 unit, thus keeping the average modulation high without creating distortion.

Antenna Matching

Because the CB radio is limited to a small amount of power output, it is important that the power be used most effectively. The power from the radio is transferred to the antenna, which converts it to electromagnetic radiation. For the most effective transfer to take place between radio and

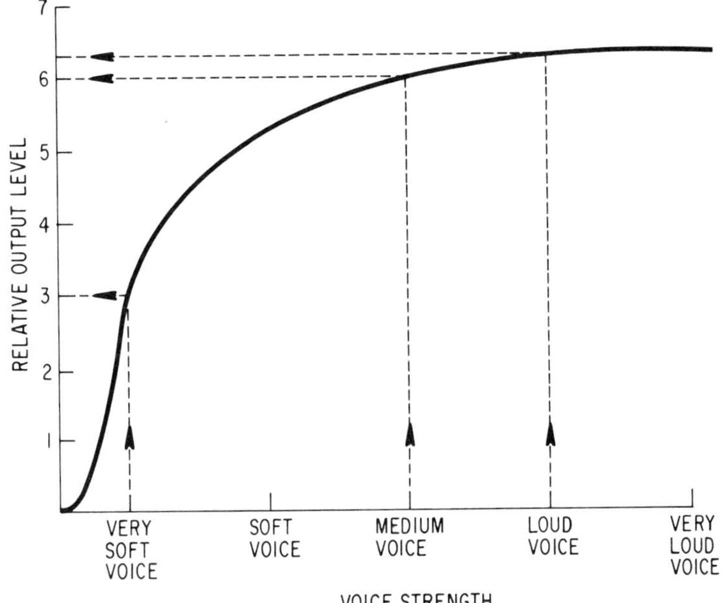

Fig. 5-2 Microphone compressor curve.

antenna, the radio output and the antenna input must be adjusted so that they are mirror images of each other. This adjustment is called *matching*. Two important conditions are involved in matching the system. One, the antenna must resonate at the frequency at which the radio is transmitting. Second, the impedance of the antenna and radio must be matched.

As explained in Chap. 4, the antenna is a conductive element related to the wavelength of the frequency in use. The more the antenna length differs from the wavelength, the greater will be the amount of energy rejected by the antenna and consequently the less energy radiated. It is therefore important to tune the antenna to the frequency to be used. If one Citizens Band channel is used the most often, the antenna may be tuned to radiate best on that channel. If all 40 channels are to be used, it is best to tune the antenna for a middle channel, such as channel 20. The antenna will normally be broad enough to do well at each end of the band as well. A mobile whip may have a moveable end section. Moving this end section in and out will lengthen and shorten the antenna. Adjusting the antenna length to match the frequency in use will achieve the greatest transfer of energy. Some antennas use capacity hats or wide elements to give them a broad band tuning ability. Whatever method is used, antennas need to be tuned for maximum performance.

Another important aspect of matching is antenna and radio impedances. Impedance refers to the resistance of a circuit which has alternating current flowing through it. This resistance is altered by the actions of components such as capacitors and coils. A capacitor is a device which has two metal surfaces separated by an insulator. Coaxial cable has a center wire which is separated from an outer tube of metal braid by plastic insulation. It presents capacitance to any AC signal traveling through it. The antenna acts like a coil designed to work best at one frequency. Because of the interaction between the coil forces of the antenna and the capacitance of the coax, impedance (resistance) increases rapidly when the antenna is not tuned to frequency and when the coaxial cable does not connect to the proper place on the antenna.

Methods have been devised to insure adequate matching. Beam antennas use gamma or "t" matching. Figure 4-5 shows a beam antenna which employs a tuning section on the driven element. This tuning section can be adjusted to permit the coax to be connected at the point which will obtain the required impedance for matching. Radio and antenna systems use an impedance of 52 ohms. They must be adjusted and matched to obtain maximum radiated power. Checks can be made using a field strength meter and a VSWR meter. The field strength meter measures radiated power and will indicate any increase in radiation achieved by tuning and matching. The VSWR meter will insure that optimum matching is achieved. The match is best when the visual standing wave ratio is lowest (closest to 1.1 to 1.0).

The *visual standing wave ratio* (VSWR or SWR) is a comparison of the amount of energy going out to the antenna and the amount of energy being reflected back from the antenna. As already indicated, a mistuned antenna or mismatched radio-to-antenna circuit causes more power to be reflected. This reflected power degrades the SWR figure, and if the SWR becomes too high, the output stage of the radio may overheat and destroy itself. The SWR accepted as optimum by most CBers is 1.1 to 1. An SWR between 1.1:1 and 2.1:1 is acceptable, but the lower value is better. An SWR of 3.1:1 indicates that one-fourth of the power is being reflected from the antenna. To lose one watt out of four cannot be tolerated.

The type and length of coaxial cable is important. Coaxial cable presents a capacitance to the radio frequency (RF) energy passing through it. This capacitance reduces the RF energy in proportion to the cable length. The shorter the cable, therefore, the less the effect of the capacitance. Many people use RG-11 instead of RG-58 cable for this reason. The greater power-handling capacity of the larger cables is not needed for CB work, but their reduced attenuation of signals makes them desirable.

Noise Reduction

A radio receiver does not distinguish between noise and sound that conveys information. The radio amplifies both. If the signal is received at about the same level as the noise, the noise will reduce the understanda-

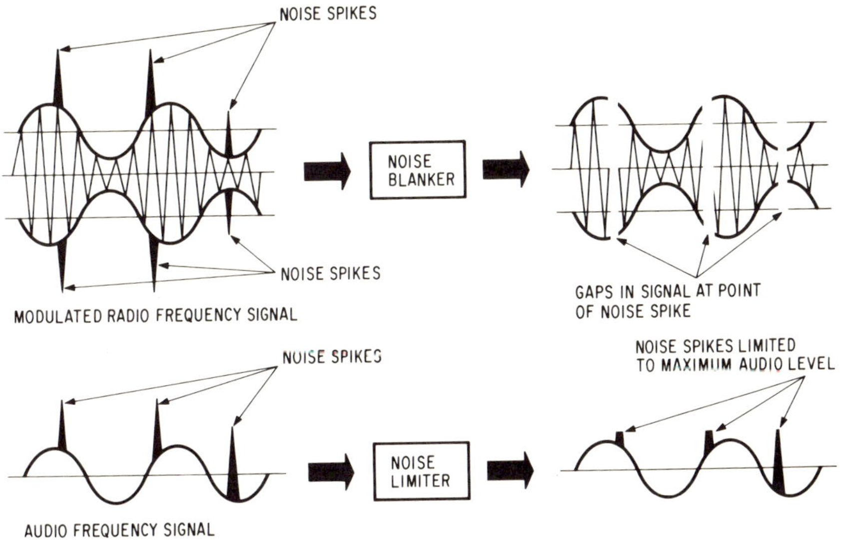

Fig. 5-3 Noise blanker and automatic noise limiter actions.

bility of the signal. A good receiver is designed to introduce as little noise as possible into the signal. Unfortunately, noise from sources outside of the radio can interfere with the signal, especially in mobile CB units.

Radios incorporate circuitry to reduce the effects of noise. One device built into many radios to help eliminate short duration noise is the *noise blanker*. It is included in the early stages before the signal is converted to audio frequency energy. Stated very simply, the action of the noise blanker is to turn the receiver off when a noise spike is detected. Since the noise spike is of very short duration, the radio can be turned off and on in much less time than is required for an audio wave to complete itself (see Fig. 5-3). The result is that the sound content will be passed, but the noise spikes will be removed.

The *noise limiter* is the second noise reduction device available in CB units. It acts on the signal after it has been converted to audio. The limiter does not turn the receiver on and off but rather limits the highest level a signal can reach. As a result, the noise spikes, which are usually higher than the signal, will be limited at a signal level equal to the highest permitted audio sound. The result is that the noise is reduced while the signal is amplified. Because noise is of short duration, it contains less average energy than does an audio signal. By limiting the spike to the audio level, the signal will sound louder than the noise because it provides more energy.

The adage, "an ounce of prevention is worth a pound of cure," is particularly true of noise. Several things can be done to reduce noise generation. At base station locations, the first requirement is to identify the noise source. One easy way to do this is to monitor the CB receiver while the possible noise sources are turned off one at a time. Once the culprit is identified, shielding, filtering, removal, repair, or replacement can be accomplished as required. A second method of finding sources of noise uses a CB walkie-talkie. A short piece of coax is attached to the unit. The built-in antenna must be disconnected from the unit input and the coax attached in its place. At the end of the coax, about 1 inch of the braid shield is removed, leaving the inner conductor and the insulating dielectric uncovered. The walkie-talkie can then be carried from place to place in the building or house. As the radio approaches a noise-producing piece of equipment, the noise will increase from the portable unit speaker. The noise source can often be found by moving the stripped end of the coax near the various parts of the equipment. The RF gain should be reduced as much as possible during this procedure.

Common sources of base station noise are fluorescent lights, motorized appliances, motorized hand tools, vacuum cleaners, and similar items. If all methods of noise reduction fail, a compromise arrangement must be reached which permits the CBer to use his radio at times when the offending equipment is not in use and vice-versa.

Noise in mobile situations is an even greater problem. The several sources of noise which must be reduced include the following:

1. Ignition systems
2. Alternators
3. Generator and electric motors
4. Turn signals
5. Voltage regulators

Noise which is generated by the ignition system (spark plugs, spark plug wires, ignition coil, and distributor) takes the form of radio frequency energy. The spark plugs ignite the gasoline in the cylinders of the automobile engine using an electric spark. Electric sparks create radio waves which act in a similar way to the transmitted waves of the CB transmitter. In fact, radio pioneers used spark gap transmitters for communication. When the ignition system of a car becomes a spark gap transmitter, the result is a distracting click which recurs with each spark. If the problem is severe, the sparks may become the prominent sound from the speaker and the desired signal may be covered up.

Curing the faulty ignition system is accomplished in several steps. Since each step increases the cost of the operation, it is best to check after each step and, if the problem is sufficiently improved, forego the remaining steps. The first step of ignition noise suppression is to clean all contacts and wires within the system. A wire brush, sand paper, and carbon tetrachloride will be helpful in this process. The spark plug wiring must be checked to be sure that the metal ends fit tightly in the distributor cap and grip the spark plug terminals well. These metal pieces can be spread or compressed slightly to improve the connection. As much oil and dirt must be removed from the spark plug wiring as possible. The wiring must be worked with carefully to keep from breaking the conductors within the cables.

Step two of ignition noise suppression is the installation of resistor type spark plugs. (Note that resistor plugs must not be used with a capacitive discharge ignition system.) Resistor plugs are usually available as an alternative in the same type as that recommended for the car. These plugs employ a resistor inside the spark plug which reduces the intensity of the spark voltage and consequently the radiation from the wiring. Many cars are factory equipped with resistor plugs. If not, a local garage or auto parts supply store will have a listing of the appropriate spark plug number. Some engine tune-up may be required after the resistor plugs have been installed.

The most radical but most efficient step for reducing ignition noise is to install shielded wiring. Spark plug cables covered with a braid shield are used, as are special metal distributor coil and spark plug shields. All these components are carefully connected and grounded to the metal frame of the car. The shielding prevents radiation of unwanted radio frequency energy by the ignition system. Again, engine tune-up may be required.

Alternator noise takes the form of a whir or a whine which changes in pitch with the speed of the engine. It can be bothersome when the engine is operating at a high rpm and the signal is weak. Filters are available which reduce or eliminate alternator noise. The filter must be chosen to match the size of the automobile load. Smaller cars with fewer accessories need smaller filters. Larger cars loaded with accessories require filters designed for more amperage. Figure 5-4 shows the proper installation method for the alternator filter. All filters should be well grounded to car frame components which do not get hot.

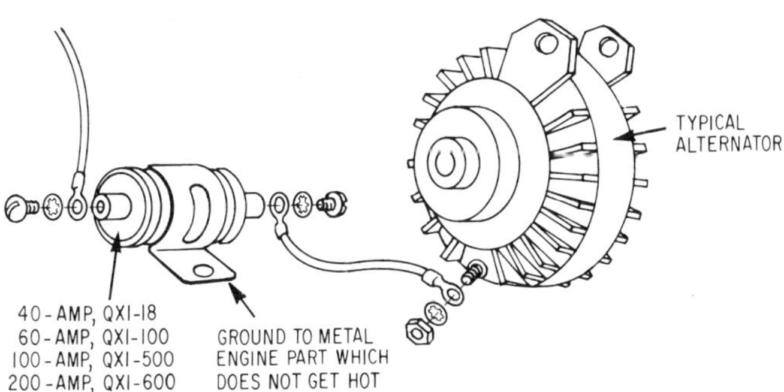

Fig. 5-4 Installation of alternator filter. (*Courtesy*, Sprague Products Co.)

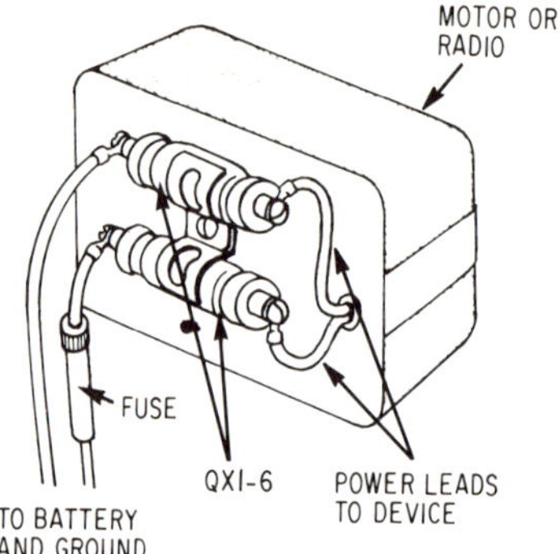

Fig. 5-5 Suppressing electric motor noise. (*Courtesy*, Sprague Products Co.)

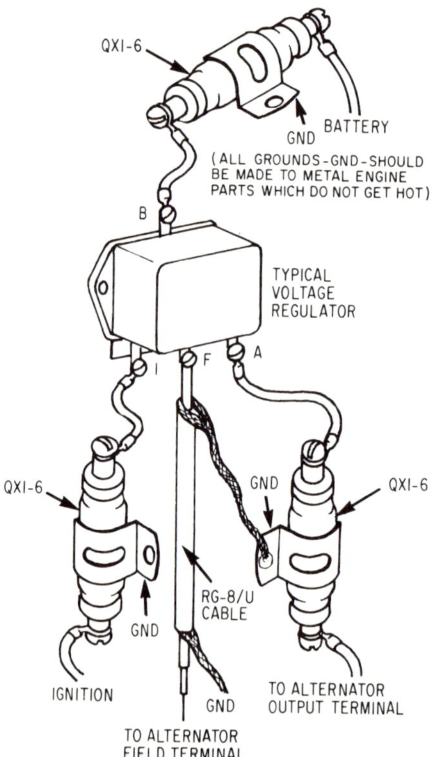

Fig. 5-6 Voltage regulator noise filtering. (*Courtesy*, Sprague Products Co.)

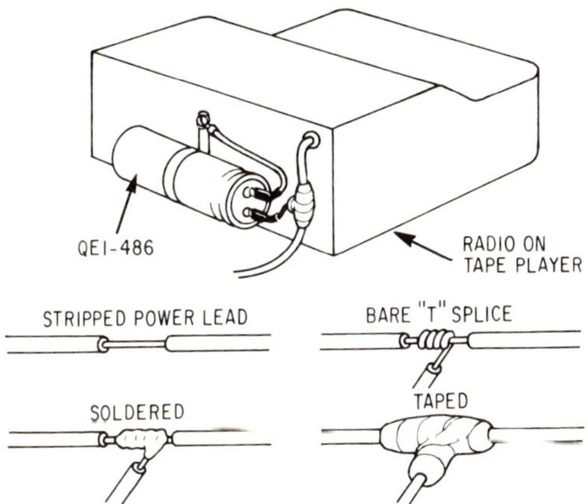

Fig. 5-7 DC Power line filtering. (*Courtesy*, Sprague Products Co.)

The electric motors used for many automobile accessories can generate noise which is heard as buzzing sounds in the CB radio. Feed-through capacitors are effective in suppressing this noise. The capacitors should be installed as indicated in Fig. 5-5.

Turn signals and voltage regulators produce sporadic clicks. These noises can be reduced by installing feed-through capacitors, as shown in Fig. 5-6. An alternative method of reducing this noise uses a filter which is tuned to reject the noise while passing the CB signals. The Gold Line model GLC 1058 is such a filter.

It is possible to install a filter on the wires supplying battery power to the CB radio. This filter will reduce noise which has not been adequately suppressed by other means. Figure 5-7 shows a typical installation method.

CHAPTER 6

Accessories for Fun and Security

If discussed comprehensively, accessories for CB radio could easily fill an entire book. The various devices described in this chapter represent a selection of the more popular; a thorough treatment has not been attempted. Accessories can be grouped into several overlapping categories. Some accessories improve the capability and convenience of CB operation. Others have little direct connection with CB radio but are popular because of the uses to which CB radio has been put. An area of growing importance deals with security. The number of CB radios and antennas stolen every day in major cities is staggering. For their own protection, CB operators must find ways of preventing CB radio theft. Many accessories are available which will assist in this task.

Increasing Capability and Convenience

Base Station Console

For many CB operators, the base station is their castle. They augment their stations with all kinds of accessories designed to make operation more efficient and convenient. Figure 6-1A shows a station console which includes a clock. The CB radio can be removed from the console for use as a mobile unit (Fig. 6-1B) and then returned to the console when base operation is desired. As indicated in Chap. 2, many base station units include a clock and speaker, as well as other features designed to make base station operation more enjoyable.

Base Station Power Supply

The CB operator may desire to use his mobile unit as a base station without investment in a console. To do so, a suitable power supply will be required, and this means the use of an ac-to-dc converter. Most mobile units operate on the 12 volts dc that is delivered by a car battery. The ac-to-dc converter changes the 110 volt ac which comes from the wall socket to 12 volts dc. A transformer changes the voltage, and a rectifier and filter circuit changes the ac to dc.

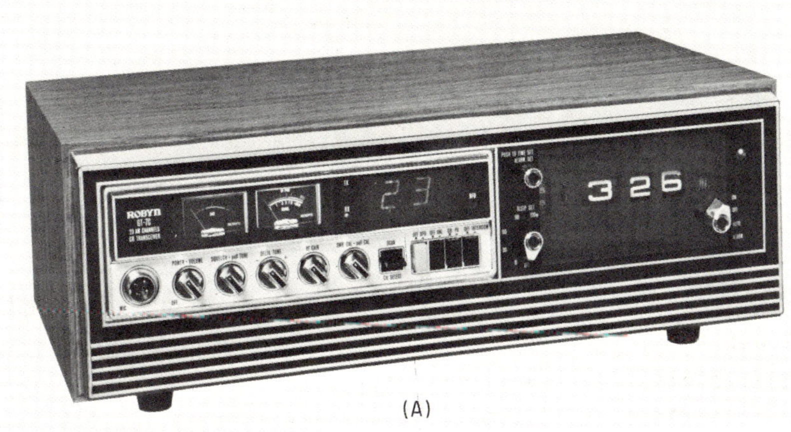

(A)

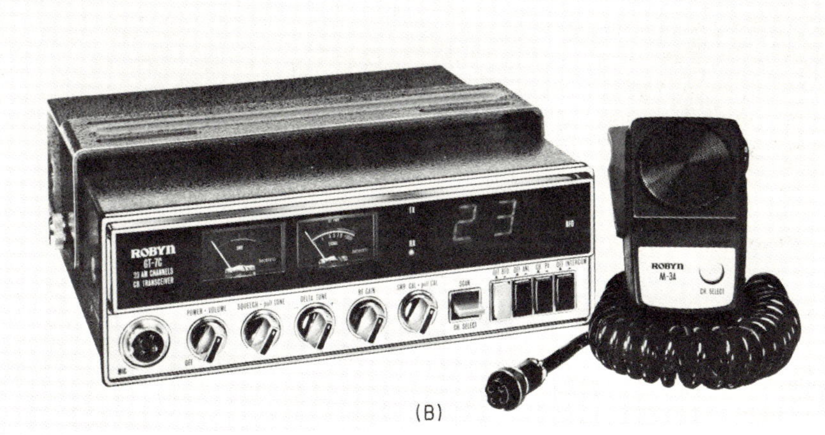

(B)

Fig. 6-1 *Base/mobile convertible:* (A) Mounted for base station use, and (B) removed from base station console for mobile use. (*Courtesy*, Robyn International, Inc.)

These converters are often called *dc power supplies* or *battery eliminators*. They are available in different types to satisfy various needs. The CB operator should choose a supply specifically designed to power CB radios. Other types of ac-to-dc converters may not be filtered well enough to prevent hum from being heard in the speaker. The amperage rating of the dc supply must be checked against the radio's requirement to make sure that it will handle the load. The power supply must be capable of supplying the greatest amount of power required. For this reason, the figure which must be matched is the load of the unit when in the transmit mode.

Fig. 6-2 *External speakers:* (A) Base station speaker, and (B) mobile speaker. (*Courtesy*, Acoustic Fiber Sound Systems, Inc.)

External Speakers

The improvement which results from the use of speakers external to the CB radio has already been discussed. For base station operation, many speakers are available. Larger speakers with better frequency response contribute to signal intelligibility. The ability to point the speakers directly at the operator also improves intelligibility since the sound is heard directly and not by some reflection path. This may be an especially important concern if the unit is placed in a console and the internal speaker is covered up or enclosed. The enclosure of the speaker is also important, as was described in Chapter 3. An open back speaker baffle can result in sound from the rear of the speaker interfering with sound from the front. Speakers like those shown in Figs. 6-2A and B use a specially designed case to enhance speaker effectiveness.

Microphones

Base station microphones are usually more elaborate than those used with mobile units (see Fig. 6-3). The push-to-talk switch is often located on the base or the stand.

There are several types of microphones, the simplest being the crystal microphone. A slab of crystalline material is mounted in the unit so that sound waves striking the diaphragm cause the crystal to twist and thereby produce an electric current that is proportional to the energy of the sound waves. The crystal microphone produces a comparatively large voltage, but it is not as rugged as other types, and its frequency response is not as good.

The most popular type is the dynamic microphone. It uses a miniature electric generator in the form of a coil of wire and a magnet to produce

Fig. 6-3 Base station microphone. (*Courtesy*, Turner Division, Conrac Corporation)

electricity from sound waves. The dynamic microphone produces less voltage than the crystal, but it is rugged and has good frequency response.

A popular type of microphone in the broadcast and recording industry is the condenser microphone. This microphone consists of two metal plates, one of which is moved by sound waves so that the space between the two is altered. This alteration causes a very small flow of electrons, which is then amplified. The result is a high quality, rugged microphone, but it must be provided with a preamplifier either within or nearby. Some CB operators adapt condenser microphones for their use, and some manufacturers are now producing condenser microphones which are attractive for the CB market.

It is important when choosing a microphone to match its output with the input of the radio. Both impedance and level must be matched. A microphone with an impedance of 150 ohms will not operate properly if connected to a radio with an input impedance of 2000 ohms. A matching transformer can be used to match the microphone to the radio. Radios can handle input signals within a range of 20 dB or more. The microphone should deliver at least the level listed for the radio.

VOX Circuits

The CB operator who wants the ultimate in convenience will require his transceiver to switch automatically between transmit and receive. A VOX or voice-operated transmit switch is the answer. The VOX responds to sound waves by closing a relay contact. The relay can be wired to operate the transmit switch on the transmitter. The result is that when the CBer begins speaking, the VOX turns the transmitter on, and when the operator ceases to talk, the VOX turns the transmitter off. For base-station operation, the VOX is a handy extra, but for mobile operation, the VOX can be a lifesaver. The mobile operator who uses a head set with a boom microphone will have both hands left free for driving, since the VOX will turn the transmitter on and off automatically.

Matching Devices

A number of accessories are available for matching the radio to the antenna. Of first importance is a means of determining when the radio and antenna are matched. A VSWR meter is useful for this purpose. The VSWR meter is placed in the antenna lead-in wire near the radio.

Some units are composed of two sections. One section is a sampler unit; the other contains the meter. The benefit of such units is that the sampler can be placed at the rear of the transceiver while the meter is placed at the front of the radio. This is a particularly helpful arrangement when one is working on a mobile unit. A meter in which the sampler and meter are both in the same box requires that the coax be brought to a convenient location, a need that cannot always be met in a mobile installation. The amount of coax which is run from the meter back to the radio must be kept as short as possible since cable length is one factor in obtaining a proper match.

The VSWR meter provides a meter reading of *power forward* (power which is actually used by the antenna) and *power reflected* (power which the antenna will not accept because of mismatch). The ratio of the reflected power to the forward power is known as the *VSWR ratio*; it is an indication of how well the antenna and radio are matched. A VSWR ratio of 1.1 to 1 is as good as can be obtained in practice. Ratios up to 2.0 to 1 are acceptable. When the ratio increases beyond this amount, much of the power is wasted, and the reflected power may be so great that it will cause the output stage of the radio to burn out.

Other types of meters are available. A field strength meter acts as a receiver which measures the amount of energy radiated by the antenna and gives a good indication of when a system is operating at peak performance. If a field strength meter reading is taken at a specific distance from the radio, this reading can be compared with other readings taken at the same distance after adjustment or equipment changes. Improvement will be indicated by a larger reading, and degradation by lesser readings.

When checking antenna systems, it is often wise to check the radio to make sure that it is operating properly. A dummy load can be used to replace the antenna. Figure 6-4 shows a simple dummy load that can be built for CB radios. Since maximum input is 4 watts, the load will not have to handle any greater amount of power than that.

Some meters include a dummy load. A meter which measures forward power only, called a *wattmeter* can be used to check radio perfor-

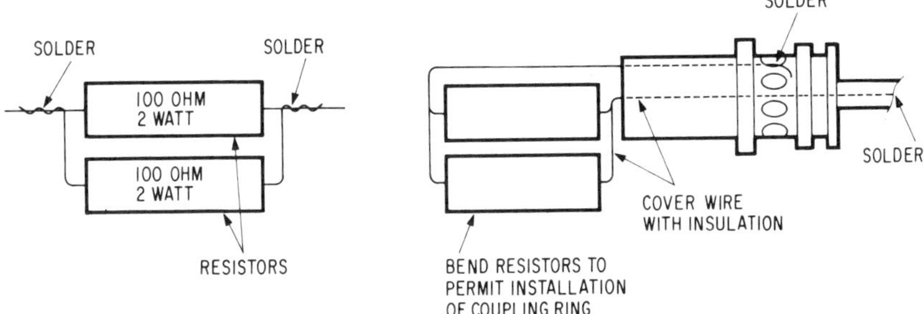

Fig. 6-4 Dummy load construction.

mance. The wattmeter can be used with a dummy load, or it can be placed in the coaxial line to measure the amount of energy going to the antenna. If it is connected with its output connection toward the radio, it will measure the amount of reflected power. Comparison of the forward and reflected readings for the dummy load with those for the antenna will give a good indication of matching.

The efficiency of antenna matching can be improved in several ways. Some antennas are tunable. The top section of the antenna can be lengthened or shortened using a set screw. Such adjustment permits the best match at the channel most often used. Antennas not capable of this tuning can be matched using a matching device. The matching device tunes the coax and antenna system for minimum VSWR. Combining a matching device with a VSWR meter will permit optimum operation at every channel.

Modulation Meter

Another meter used by CBers is the modulation meter. This meter indicates the percentage of modulation of the CB transmitter. If the modulation is low, adjustments can be made or microphone preamplifiers added to increase it. In some cases it may be necessary to take the unit to a repair shop for alignment and adjustment.

Modulation meters are often included in multipurpose meters called *CB testers*. Since these meters perform a variety of functions, they can be an economical choice. The CBer must bear in mind that adjustment of frequency and the power-determining parts of the radio is restricted to FCC-licensed technicians.

Filters

CB radios possess the potential for interfering with nearby television sets. This phenomenon is called *TVI*. If TVI is bothersome, a high-pass filter can be installed on the TV set to reduce it. This filter permits high-frequency signals to pass through but rejects low frequencies, including CB signals. In a similar fashion, low-pass filters can be installed on the radio. This filter permits the lower frequency CB signals to pass through to the antenna while rejecting higher frequencies. Some CB radios produce harmonics of the basic CB frequencies. A *harmonic* is a frequency that is a multiple of the desired frequency. If these harmonics are too strong, they can interfere with other radio services. The low-pass filter prevents their transmission.

Miscellaneous Accessories

Some accessories commonly used by CBers have no direct link to CB radio. One prevalent use of CB radio is sharing information about police locations. A radar detector is one means of discovering a traffic patrol's location. Radar detectors like the ones shown in Fig. 6-5 are receivers

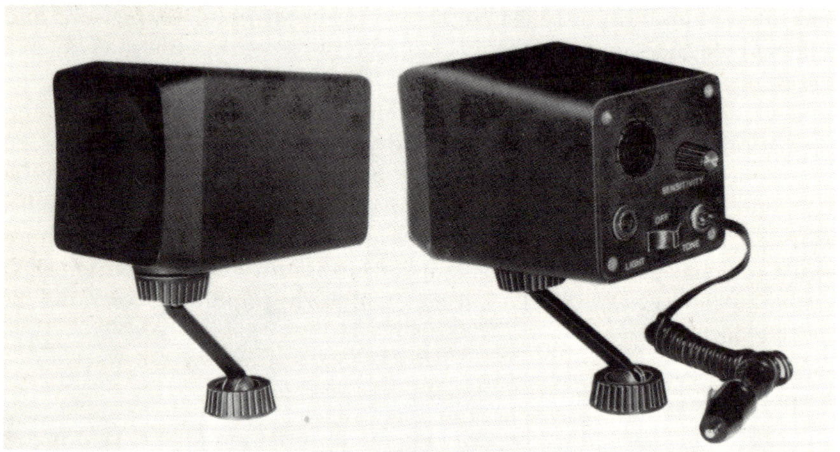

Fig. 6-5 Radar detectors. (*Courtesy*, Universal Machine Co., Inc.)

which are designed to produce an audible signal when the pulses of a traffic radar are detected. Since they are not complicated in construction, they are relatively inexpensive. Contrary to rumor, moreover, they are not illegal. Only devices which interfere with the proper operation of traffic radar systems are illegal, and radar detectors *receive* radar pulses only. Some users praise them loud and long. Others are more conservative in their approval. There is little doubt that they can prove helpful in many situations. It is also true, however, that an individual significantly exceeding the speed limit will probably be clocked by the police before he can slow down despite his advance warning. For the cautious driver who wants an inexpensive reminder of his speed, the radar detector is a worthwhile investment.

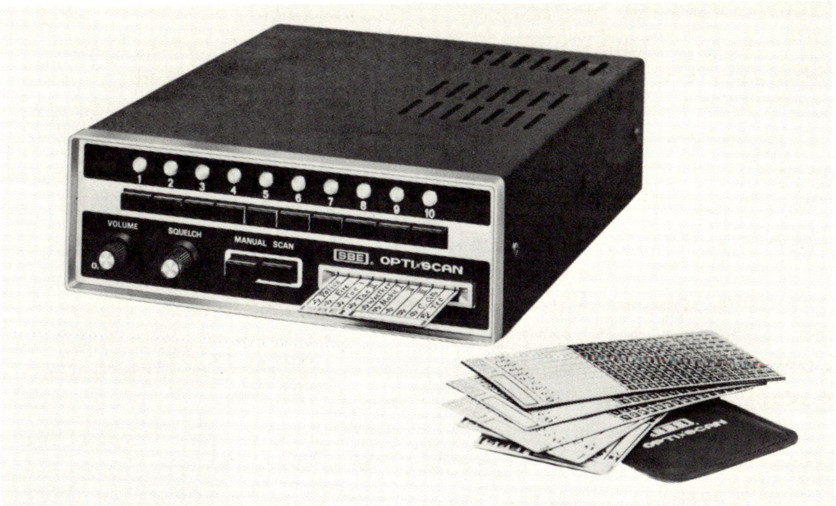

Fig. 6-6 Scanning monitor-receiver. (*Courtesy*, SBE Linear Systems, Inc.)

Many CBers employ their radios in public assistance endeavors. Some are members of Civil Defense teams. Many belong to groups such as ALERT and REACT. A lot of CBers seek opportunities to assist motorists and others on an individual basis. If a CBer serves in several of these capacities, he may find himself obligated to monitor numerous channels simultaneously. Purchasing a separate receiver for each channel can irritate the XYL or OM "real quick" (XYL is CB slang for wife; OM is the male counterpart). An accessory which answers the dilemma is the scanning monitor, which is capable of receiving several channels simultaneously. It does so by sampling each channel one after another. Only when a channel is in use will the scanner stop and connect it to the speaker. When the channel is clear of traffic, the scanning is resumed. Some scanners scan only a certain number of channels but do so continuously. The unit shown in Fig. 6-6 is programmable, permitting the scanned channels to be chosen for specific requirements and then reprogrammed for other requirements. Some scanners allow for priority channels. The monitor will respond to a priority channel whenever traffic is on it, even if another channel earlier in the scanning sequence has traffic on it. Some scanners even permit mixing of high and low frequency signals in the scanning sequence. Scanners can be obtained for police and industrial frequencies as well as CB frequencies.

Weather monitors and other special purpose radios are popular with CBers. Many consider it a part of their duty to pass on the latest weather report or WWV (Government Bureau of Standards radio service) time check.

Accessories for Security

The phenomenal increase in the number of CB enthusiasts has been coupled with an equally phenomenal number of CB thefts. In many cities and larger communities, hardly an hour passes without a report of a stolen CB unit. The problem has become so widespread that a new CB market has been created—accessories for CB security.

CB theft usually involves mobile equipment. Thieves prey on vehicles left in isolated or unwatched areas. These thieves are often highly professional. They can spot a likely car and lift the CB radio in the space of a very few minutes. Antenna theft takes even less time.

Vehicles which are easy pickings for CB thieves are those with triangular side windows since these can be readily forced open. Pickup trucks are reportedly easy to break into as well.

If difficulty is encountered in removing the unit, the thief often resorts to a crowbar. Needless to say, a crowbar can ruin the appearance of a car's dash.

Alarm Systems

Alarm systems have been developed for CB security which often frighten the thief away before he can remove the radio or use a crowbar. Several types of alarms are available, each of which operates in a different way. One type is activated when a unit is removed from its normal position. A loud sound is generated which is intended to frighten the thief and alert the authorities. Some systems are designed to be activated by entry into the vehicle. This type of alarm is better because it is activated before the unit can be touched. There is less likelihood of the would-be thief making good his escape with the radio if the alarm goes off when he first gains access to the car. This type of alarm system obviously provides protection against car theft as well as CB radio theft.

Another type of alarm, which is somewhat ingenious in design, is placed in the RF coax line. It senses when the coax is disconnected from the radio or antenna or when it is cut and responds with a loud audio signal. The beauty of this system is twofold. For one thing, the alarm unit itself is more secure since it can be mounted in the trunk (if a trunk or rear mounted antenna is used). Moreover, it provides fewer clues to its presence. Coax alarms protect both radio and antenna.

There are problems which are common to all alarm systems. A major one is that they all depend on someone being alerted to the robbery. If no one is in the vicinity of the car or if other sirens and sounds confuse the alarm signal, the alarm may go unheeded.

Most currently available alarms utilize an external speaker and power from the car battery. If a thief suspects that an alarm system is present, all he has to do is cut the leads to the system speaker or the car battery before removing the radio. The ideal alarm system would utilize its own power source and built-in speaker. It would be constructed so that neither could be reached without setting off the alarm. Ideally, an alarm would be tripped by tampering prior to the actual removal of a unit. Sensing devices are available which respond to different stimuli. A human touch will set off some alarms. Vibrations will trigger others. Use of such devices would require greater ingenuity on the part of the would-be thief. The important requirement is that the power supply and sounding device be inaccessible from the outside of the unit. Of course, the alarm must have disarming provisions available to the owner.

Mobile Mounts

Special mounting devices have been developed which discourage CB theft. One such device combines an alarm system with a locking mount. When the CB is placed in the mount and locked in place, it is protected in two ways. First, to remove the radio, the mount must be broken open. This is in itself difficult, and will discourage the amateur thief. Second, when the mount is broken open, the alarm system is activated. If the thief does not do a good job of breaking the mount open on the first try, the alarm may dissuade him from a second attempt. Such a system will limit theft to more isolated areas where rapid flight is easier.

A number of locking mounts are available which do not employ alarms. They take several forms. Essentials for locking mount effectiveness include mechanical strength, secure mounting methods, secure locking methods, and freedom from lock defeat.

The first of these essentials, mechanical strength, is at the heart of the locking mount's claim to fame. If the mount can be broken easily, it will provide little protection. Cast metal parts should be looked at with suspicion. If they are brittle, they may be easily broken off. Metal which bends is acceptable, providing that the mount fits the radio snuggly. If it does, bending the parts of the mount will not free the unit.

The mount can be strong, and the radio may still not be safe. Just as a door which has screws and hinges on its exposed exterior can be easily opened by simply removing the screws, a locking mount whose mounting screws are accessible is not secure. For ultimate security, the mount should be affixed to the dash with bolts and self-locking nuts. The head of the bolt should be inaccessible when the CB radio is in place. Since the locking nut will loosen only when the bolt is held from the bottom, it will not be possible to remove the nut. Its location under the dash will not make the attempted theft any easier either. When the nut is loosened far enough, the bolt will begin to turn, and the mount will be secure. It is wise to install large washers between the dash and the nut. If a thief attempts to pry the unit out of the car, there will then be less likelihood of the nuts being pulled through the metal of the dash.

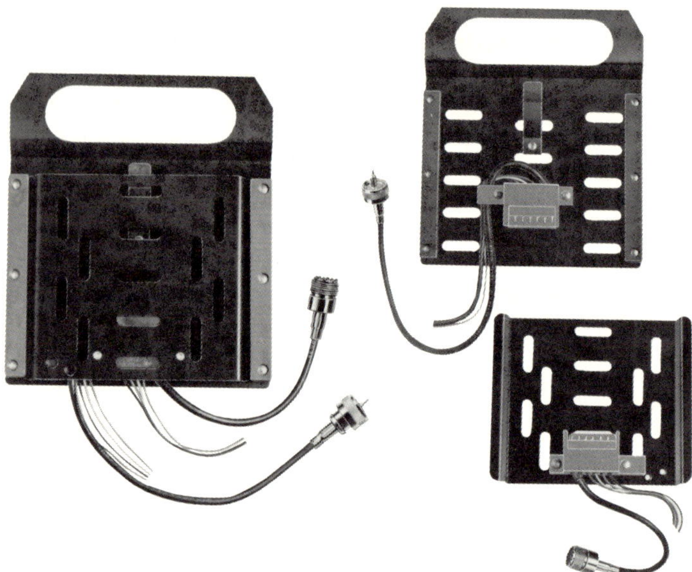

Fig. 6-7 Slide mount. (*Courtesy*, Universal Machine Co., Inc.)

To be secure, the lock of the mount must be tamperproof. The CBer should check the locking mechanism prior to purchase to be sure that it is not easily defeated. It is also wise to check the method used to hold the radio in the mount to be sure that it cannot be defeated.

It should be understood that most antitheft devices are expected to be little more than deterrents to crime. Professionals will pick locks or devise other ways to defeat new devices as quickly as they are developed. Amateur thieves will use more clumsy methods and may damage both radio and car even when they fail to remove the former. Insurance is always a wise investment.

A slide mount (Fig. 6-7) is one way of preventing radio theft. This mount permits a radio to be slid into it and to be automatically connected to power, speaker, and antenna once it is in place. This is usually accomplished by attaching a special connector plate to the radio which will mesh with a second connector plate attached to the car dash. Connections are made between the automobile systems and the car-mounted plate and between the radio and its plate. As soon as the radio is inserted, it is ready to operate. When it is removed, no cable and wire have to be detached. The benefit of the device is that an operator can carry the radio with him when he leaves his vehicle. When he returns to the car, he simply slides the radio in and starts operating. Unless the CBer is mugged, his radio should be safe. Even if he is mugged, the radio might prove to be a useful weapon.

Some recent CB radios are designed for custom mounting in the dash. The unit shown in Fig. 6-8 combines the CB function with standard AM and FM radio capability. This type of CB radio replaces the existing car radio and provides security because of the complication of the in-dash mounting. Removing the radio is as difficult as putting it in, and thieves will think twice before trying to do so. The result of using this type of CB transceiver is a relatively secure unit with an attractive, finished appearance.

It is possible to install CB units in inconspicuous locations. Some CBers place them in the glove compartment. With the compartment door closed and locked, the presence of the unit is concealed. If such an installation is coupled with an antenna which does not advertise the presence of a CB radio, thieves may be fooled. It should be noted that such mounting reduces operating convenience since the operator must reach across the car to operate the set.

Operating convenience is a problem with hidden mounting wherever the unit is located. The unit could be placed under the front seat, but reaching the channel selector and other dials would be difficult. If the unit is stolen, however, the knobs will not be reachable at all. In these kinds of installations, the microphone also must be hidden from view. A mike connected to a wire running under the seat is a sure giveaway.

Some recent transceivers have reduced the problem of the hidden mounting. A number of units allow some of their functions to be controlled on the microphone. The CBer can change channels, adjust micro-

Fig. 6-8 Dash-mounted CB radio with AM and FM provisions. (*Courtesy*, J.I.L. Corporation of America)

Fig. 6-9 Remote control CB radio. (*Courtesy*, Royce Electronics Corporation)

phone and speaker levels, and perform other control functions on the microphone alone. This arrangement is convenient for driving and also makes mounting of the transceiver in an inaccessible place more feasible.

The ultimate in hidden transceivers is the completely remote-controlled unit, as shown in Fig. 6-9. Patterned after commercial two-way radios, this unit uses a control head that is very small (comparable in size to a pack of cigarettes, according to the advertisement). The control head contains a minimum of electronics since these are contained in a separate section designed for mounting in the trunk or under the seat. This separate section can be made secure, and the control head can be made as inconspicuous as possible. Such radios obviously solve many problems.

Antennas

Police report that CB thieves drive about parking lots in search of CB radios to steal. It is the CB antenna that alerts them to vehicles containing one. It stands to reason that if the antenna were made less noticeable, CB thieves would have more difficulty in spotting their marks. Several antenna designs and devices assist in such camouflage.

CB antennas can be obtained which mount in the same way as, and have the appearance of, standard fender-mounted antennas. Unless a loading coil is built into such an antenna, loading will be accomplished at the bottom of the antenna. This will reduce efficiency, especially if the

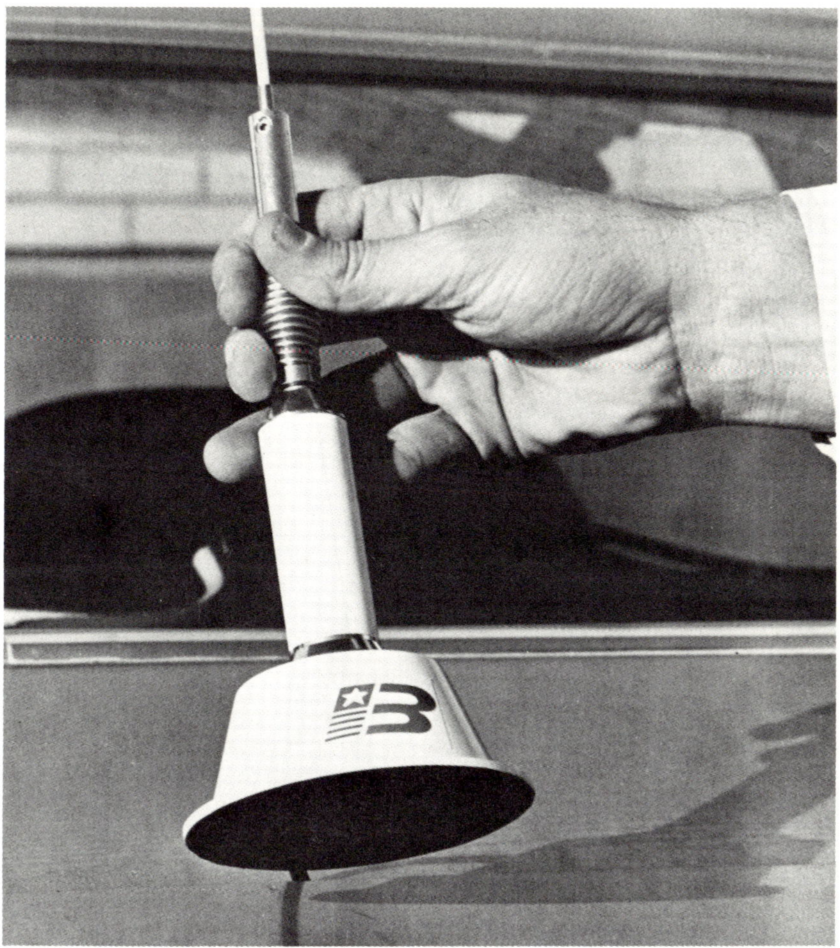

Fig. 6-10 Magnetic base antenna. (*Courtesy*, Breaker Corporation)

loading is accomplished under the fender, but the security benefit of this type of antenna may make it worthwhile even if it does not radiate at maximum levels. This antenna's camouflage lies in the fact that it looks just like standard radio antennas. The prospective thief will have to inspect every car with a standard antenna to find the ones which have CB units. Seldom will they do this.

Another antenna which is useful for security purposes is one with a magnetic mount. This antenna mount incorporates a strong magnet which adheres to the metal surface of the vehicle (see Fig. 6-10). The magnetic bases of these antennas are designed to hold very tightly and will remain in place even when driving at the speed limit. The antenna is not designed to be a permanent fixture. For security purposes, it cannot be. On the contrary, it is meant to be removed when the vehicle is stopped and can be easily replaced prior to resuming motion. Removing the antenna and storing it inside the vehicle presents passing CB snatchers with not one clue.

The magnetic mount has some drawbacks. The cable must be mounted to the outside each time the antenna is removed and replaced. This job can be a hassle. Also, the routing of the cable can be a problem. To run it through a window interferes with the opening and closing of the door. It can be run out around the door and the door closed on it. If the door and frame fit too closely, however, the cable will be crimped, and VSWR may increase. Even with these problems, however, the magnetic mount antenna is a useful device.

Another antenna for security purposes is the gutter-mount clip-on antenna, which clips on the rain gutter of the car. Installation is quick and easy, but this type of antenna has the same cable routing problems as the magnetic mount. Moreover, it is limited to installation at the side of the vehicle. This is not as desirable as the center roof location which the magnetic mount can utilize. At the same time, however, since the gutter-mounted antenna is located near the side of the vehicle, less cable flap may be experienced when in motion than might be with a magnetic mount placed further in on the roof. Security is enhanced by the easy removal of the antenna.

An ingenious mount which has recently been made available is very useful for cars with trunks. It is designed to attach to the lip of the luggage compartment lid. When the lid is closed, the mount protrudes through the space between the lid and the car body. When the lid is raised, the mount permits the antenna to be lowered into the luggage compartment. When the lid is then closed, no trace of the antenna or mount is observable. The benefit of such a mount is easily seen.

Another type of antenna mount is similar in principle to the one described above (see Fig. 6-11). This device uses the normal trunk lip mount antenna base. A special latch connects to the base and pulls it tight against the trunk lid lip. Set screws are not used. When the CBer leaves his vehicle, he simply unlatches his antenna and lays it in the trunk. When he is ready to go again, he places the mount on the trunk lid lip and latches it down. The process is simple.

While on the subject of security, a few precautionary procedures should be discussed. Some of the alarms described earlier provide a decal for display in the CBer's window. The idea is that a potential thief may decide not to try to steal the unit if he is alerted to the alarm. If the ruse works, the CBer can save a lot of money on damages resulting from the removal of the set with hammer and crowbar.

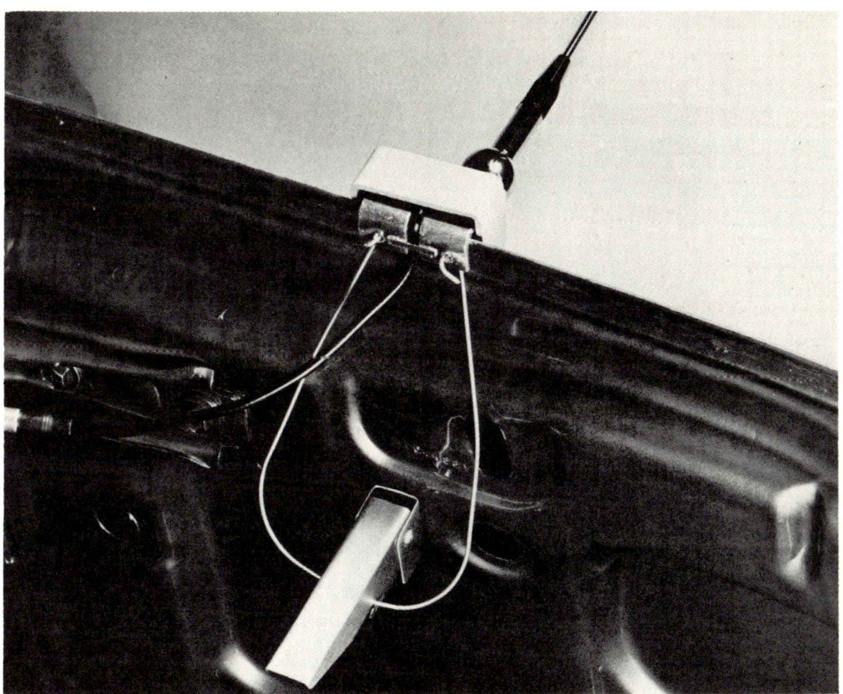

Fig. 6-11 Trunk lid mount antenna latch. (*Courtesy*, South Com, Inc.)

Another important deterrent to CB theft is registration. Many police departments register CB radios. This practice facilitates the return of the units to their rightful owners if they are recovered. Some local and national organizations catalog serial numbers for subsequent use. FCC regulations require serial numbers to be indelibly inscribed on the chassis of CB units. This helps to discourage theft.

One last procedure is also very helpful. The CBer should engrave his name and address on the radio. To do so will aid in recovery and may help deter thieves since the information will be indelibly inscribed. It is difficult to sell equipment inscribed with someone else's name and address on it.

APPENDIX

CB SLANG

Advertising	A police car with flashing lights on
Affirmatory	Yes
Back door	Vehicle at rear of a group of CB-interconnected vehicles
Back out	Leave a conversation
Backstroke	Return trip
Bagging	Police catching speeders
Bang-a-U'y	Make a U-turn
Barefoot	Operation of CB unit without illegal amplifier
Bean store	Cafe
Bear	Police
Bear bait	Car without CB radio
Bear bite	Speeding ticket
Bear cage	Police station
Bears in the air	Police in airplanes or helicopters
Beat the bushes	Driving fast enough to lure police from hiding without getting a ticket
Beaver	Female
Big Daddy	FCC
Big skip land	Heaven
Big switch	On/off switch on CB radio
Big 10-4	Agreement
Bit on the britches	Received speeding ticket
Black and white	Police car
Bodacious	Strong (signal)
Bootleg	Operating with a stolen call
Box	CB radio
Break	Request to enter or interrupt a CB interchange
Breaking up	Signal becoming lost in interference
Brown paper bag	Unmarked police car
Brush your teeth and comb your hair	Radar ahead
Bubble machine	Police car
Bucket mouth	CBer who talks too much
Camera	Radar
Candy man	FCC
Charlie (Uncle Charlie)	FCC
Check the seat covers	Look at passengers
Chicken coop	Truck weighing station
City kitty	Town police
Clean	No police in view
Comeback	Response to a call
County mountie	Sheriff or deputy
Covered up	Interfered with by other signals
Cut some Z's	Sleep
Double nickel	Driving 55 miles per hour
Drop the hammer	Drive fast
Ears	CB radio

Eighteen wheeler	Multiwheeled truck
Eighty-eights (88's)	Love and kisses
Eyeball	Face-to-face meeting
Eye in the sky	Police aircraft
Fed	FCC inspector
Feed the bears	Pay a fine for speeding
Final	Last transmission of a contact
Fix	Location
Flaps down	Slow or slowing
Flip flop	Return trip
Foot warmer	Illegal amplifier
Forty Roger	Understood, and reply is affirmative
Four wheeler	Car
Fox Charlie Charlie	FCC
Front door	First vehicle in a group of CB-linked vehicles
Gear jammer	Truck driver
Getting out	Transmitting well
Give me a shout	Call me
Good buddy	General term used by one CBer in addressing another
Good numbers	73's (best wishes) and 88's (love and kisses)
Got a copy?	Do you hear?
Green stamps	Money
Hammer	Accelerator
Hammer down	Going fast
Hammer up	Slowing down
Handle	Name used on CB radio
Have a safe and sound one	Best wishes
Home 20	Home location (modification of 10-20)
Jaw Jacking	Talking, shooting the breeze
Kiddie car	School bus
Kodiak with a Kodak	Radar Unit
Land line	Telephone
Linear	Illegal amplifier
Load of post holes	Truck without a load
Load of VW radiators	Truck without a load
Local yokel	City or town police
Make a trip	Move to alternate channel
Mercy	A nonsense word when used by CBers
Modulate	Talk
Motion lotion	Gasoline or diesel fuel
Mountie	County police officer
Negative contact	Unable to establish contact
Negative copy	Inability to hear and understand
Negatory	Negative reply, no
One more time	Transmit a final
On the side	Listening, not transmitting
On the peg	Driving the speed limit
Over shoulder	To the rear
Paper hanger	Police giving tickets
Pass the numbers	Sign off (73's & 88's)
Peanut butter in the ears	Operator not listening; CB turned off
Picture taker	Radar
Plain brown wrapper	Unmarked police car
Portable barn yard	Cattle truck
QSL, QSL card	A post card verifying a contact between CBers
Raise	Call a station
Ratchet jaw	Talk, shoot the breeze
Read	Hear

Reading the mail	Listening
Rig	CB radio
Rock	Crystal used to determine frequency of radio
Rocking chair	Vehicle in center of CB-linked group
Roger	Received and understood
Scatter stick	Antenna
Seventy-three's (73's)	Best wishes
Shake the trees and rake the leaves	In a group of CB-linked vehicles, lead vehicle looks for police ahead and rear vehicle watches behind
Short short	Soon
Shout	Call on the radio
Shoveling coal	Speeding up
Slammer	Jail, prison
Smokey	Police
Smokey report	Report on location of radar unit
Smokey dozing	Police car not in motion
Smokey on rubber	Police car in motion
Smokey with ears	Police car equipped with CB radio
Smokey on the ground	Policeman out of his car
Thermos bottle	Tanker truck
Tijuana taxi	Identifiable police car with light and markings
Turkey farm	Rest area
Two-wheeler	Motorcycle
Uncle Charlie	FCC
Walking on you	Interfering with you
Wallpaper	QSL card
Wall-to-wall and tree-top-tall	Indicates that signal is very loud and clear
Water hole	Truck stop
Willy weaver	Drunk driver
Work twenty	Place of employment (Modification of 10-20)
X-ray machine	Radar
XYL	Wife
YL	Young lady

Index